P9-CAM-952

WH
547-068

174843

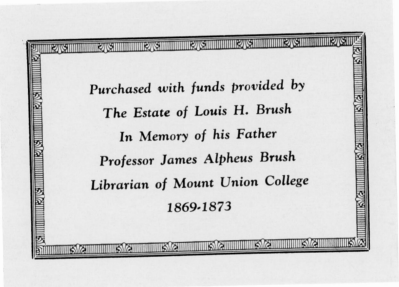

Purchased with funds provided by
The Estate of Louis H. Brush
In Memory of his Father
Professor James Alpheus Brush
Librarian of Mount Union College
1869-1873

ORGANIC SYNTHESES

ORGANIC SYNTHESES

AN ANNUAL PUBLICATION OF SATISFACTORY
METHODS FOR THE PREPARATION
OF ORGANIC CHEMICALS

VOLUME 55

1976

ADVISORY BOARD

C. F. H. ALLEN
RICHARD T. ARNOLD
HENRY E. BAUMGARTEN
RICHARD E. BENSON
A. H. BLATT
VIRGIL BOEKELHEIDE
RONALD BRESLOW
T. L. CAIRNS
JAMES CASON
J. B. CONANT
E. J. COREY
WILLIAM G. DAUBEN
WILLIAM D. EMMONS
ALBERT ESCHENMOSER
L. F. FIESER
R. C. FUSON
HENRY GILMAN
W. W. HARTMAN
E. C. HORNING

HERBERT O. HOUSE
JOHN R. JOHNSON
WILLIAM S. JOHNSON
N. J. LEONARD
B. C. McKUSICK
C. S. MARVEL
MELVIN S. NEWMAN
C. R. NOLLER
W. E. PARHAM
CHARLES C. PRICE
NORMAN RABJOHN
JOHN D. ROBERTS
R. S. SCHREIBER
JOHN C. SHEEHAN
RALPH L. SHRINER
H. R. SNYDER
MAX TISHLER
KENNETH B. WIBERG
PETER YATES

BOARD OF EDITORS

SATORU MASAMUNE, *Editor-in-Chief*

ARNOLD BROSSI
GEORGE H. BÜCHI
ROBERT M. COATES
ROBERT E. IRELAND

CARL R. JOHNSON
WATARU NAGATA
WILLIAM A. SHEPPARD
ZDENEK VALENTA

WAYLAND E. NOLAND, *Secretary to the Board*
University of Minnesota, Minneapolis, Minnesota

FORMER MEMBERS OF THE BOARD, NOW DECEASED

ROGER ADAMS
HOMER ADKINS
WERNER E. BACHMANN
WALLACE H. CAROTHERS
H. T. CLARKE
ARTHUR C. COPE

NATHAN L. DRAKE
CLIFF S. HAMILTON
OLIVER KAMM
LEE IRVIN SMITH
FRANK C. WHITMORE

JOHN WILEY AND SONS
NEW YORK · LONDON · SYDNEY · TORONTO

Copyright © 1976 by John Wiley & Sons. Inc.

All rights reserved. Published simultaneously in Canada.

No part of this book may be reproduced by any means, nor transmitted, nor translated into a machine language without the written permission of the publisher.

"John Wiley & Sons, Inc. is pleased to publish this volume of Organic Syntheses on behalf of Organic Syntheses, Inc. Although Organic Syntheses, Inc. has assured us that each preparation contained in this volume has been checked in an independent laboratory and that any hazards that were uncovered are clearly set forth in the write-up of each preparation, John Wiley & Sons, Inc. does not warrant the preparations against any safety hazards and assumes no liability with respect to the use of the preparations."

Library of Congress Catalog Card Number: 21-17747
ISBN 0-471-57390-6

Printed in the United States of America

10   9   8   7   6   5   4   3   2   1

547
068
174843

NOMENCLATURE

Common names of the compounds are used throughout this volume. Preparations appear in the alphabetical order of common names of the compound or names of the synthetic procedures. The *Chemical Abstracts* indexing name for each title compound, if it differs from the common name, is given as a subtitle. Because of the major shift to new systematic nomenclature adopted by *Chemical Abstracts* in 1972, many common names used in the text are immediately followed by the bracketed, new names. Whenever two names are concurrently in use, the correct *Chemical Abstracts* name is adopted. The prefix *n*- is deleted from *n*-alkanes and *n*-alkyls. In the case of amines, both the common and systematic names are used, depending on which one the Editor-in-Chief feels is more appropriate. All reported dimensions are now expressed in Système International units.

SUBMISSION OF PREPARATIONS

Chemists are invited to submit for publication in *Organic Syntheses* procedures for the preparation of compounds that are of general interest, as well as procedures that illustrate synthetic methods of general utility. It is fundamental to the usefulness of *Organic Syntheses* that submitted procedures represent optimum conditions, and the procedures should have been checked carefully by the submitters, not only for yield and physical properties of the products, but also for any hazards that may be involved. Full details of all manipulations should be described, and the range of yield should be reported rather than the maximum yield obtainable by an operator who has had considerable experience with the preparation. For each solid product the melting-point range should be reported, and for each liquid product the range of boiling point and refractive index should be included. In most instances it is desirable to include additional physical properties of the product, such as ultraviolet, infrared, mass, or nuclear magnetic resonance spectra, and criteria of purity such as gas chromatographic data. In the event that any of the reactants are not commercially available at reasonable cost, their preparation should be described in as complete detail and in the same manner as the preparation of the product of major interest. The sources of the reactants should be described in notes, and physical properties such as boiling point, index of refraction 5 and melting point of the reactants should be included except where standard commercial grades are specified.

Beginning with Volume 49, Method of Preparation (Sec. 3) and Merits of the Preparation (Sec. 4) have been combined into Discussion (Sec. 3). In this section should be described other practical methods for accomplishing the purpose of the procedure that have appeared in the literature. It is unnecessary to mention methods that have been published but are of no practical synthetic value. Those features of the procedure that recommend it for publication in *Organic Syntheses* should be cited (synthetic method of considerable scope, specific compound of interest not likely to be made available commercially, method that gives better yield or is less laborious than other methods, etc.). If possible, a brief discussion of the scope and limitations of the procedure as applied to other examples, as well as a comparison of the particular method with the other methods cited, should be included. If necessary to the understanding or use of the method for related syntheses, a brief discussion of the mechanism may be placed in this section. The present emphasis of *Organic Syntheses* is on model procedures rather than on specific compounds (although the latter are still welcomed), and the Discussion should be written to help the reader decide whether and how to use the procedure in his own research. Three copies of each procedure should be submitted to the Secretary of the Editorial Board. It is sometimes helpful to the Board if there is an accompanying letter setting forth the features of the preparations that are of interest.

Additions, corrections, and improvements to the preparations previously published are welcomed and should be directed to the Secretary.

CLIFF S. HAMILTON

November 23, 1889–April 7, 1975

Cliff S. Hamilton, Editor-in-Chief of Volume 29 of *Organic Syntheses*, who joined the Board of Editors in 1942 and was a long-time member of the Advisory Board, passed away on April 7, 1975 in Denver, Colorado at the age of 85.

Born in Blair, Nebraska on November 23, 1889, he received his B.S. degree from Monmouth College in Illinois in 1912. He was a student at the University of Illinois during 1914–1915 and served in the Chemical Warfare Service in 1917 during World War I. Then he served as an instructor in chemistry at Ohio Wesleyan University from 1917 to 1919. He began graduate work at the University of Minnesota during 1919–1920. When Lauder Jones, with whom he worked, moved to Princeton, Hamilton went to Northwestern University to work with W. Lee Lewis, and received his Ph.D. degree there in 1922. After a year as a research instructor in pharmacology at the University of Wisconsin in 1922–1923, he returned to his native state and joined the faculty of the University of Nebraska where he served as an assistant and associate professor from 1923 to 1927. From 1927 to 1929 he served as an associate professor at his graduate *alma mater*, Northwestern University. In 1929 he returned to Nebraska as a full professor of chemistry and served as chairman of his department from 1939 to 1955, and also as dean of the graduate college during 1938–39 and 1940–41, retiring in 1957. He was a long-time consultant to Parke, Davis and Company in Detroit from 1927 until 1963. He was Chairman of the Nebraska Section in 1924–26, Chairman of the Organic Division in 1940, and Chairman of the Divisional Officers Group of the American Chemical Society in 1940–41. He served as Associate Editor of *Organic Reactions, Volume 2*, in 1944, and as an Associate Editor of *Chemical Reviews* from 1946 to 1948. His honors included an honorary Doctor of Science degree from his undergraduate *alma mater*, Monmouth College, in 1954, the Midwest Award of the St. Louis Section of the American Chemical Society in 1955, and the dedication of the new chemistry building at the University of Nebraska in his honor on October 25, 1970. His research interests included the synthesis of organic compounds containing arsenic, antimony, or phosphorus, and the study of heterocyclic compounds utilizable as drugs. He helped develop Mapharsen, an arsenical formerly widely used against syphilis, and Camoquin, an antimalarial. He is survived by his wife, Frances Howe Hamilton, three children, Robert W. (a research chemist with G. D. Searle and Co., who resides in Wilmette, Ill.), Martha H. (Mrs. Amos W. Dickey, Denver, Colo.), Dr. Clif S. (a thoracic and cardiovascular surgeon in Fargo, N. D.), and five grandchildren.

We shall miss this valued friend, adviser, and contributor to the success of *Organic Syntheses*.

Wayland E. Noland

PREFACE

Recent volumes of *Organic Syntheses* have laid emphasis on widely applicable, model procedures that illustrate important types of reactions. This volume continues this policy, and many of the procedures selected here have major significance in the synthetic method, rather than in the product that results. However preparations of reagents and products of special interest are also included, as in previous volumes.

The selection of the thirty procedures clearly reflects the current interest of synthetic organic chemistry. Thus seven of them illustrate uses of Tl(I), Tl(III), Cu(I), and Li(I), and three examples elaborate on the process now termed "phase-transfer catalysis." In addition, newly developed methods involving fragmentation, sulfide contraction, and synthetically useful free radical cyclization are covered in five procedures. Inclusion of preparations and uses of five theoretically interesting compounds demonstrates the rapid expansion of this particular area in recent years and will render these compounds more readily and consistently available.

Introduced into *Organic Syntheses*, presumably for the first time, are Tl(III) salts that achieve simple preparations of METHYL 2-ALKYNOATES FROM 3-ALKYL-2-PYRAZOLIN-5-ONES and also effect regioselective iodination to obtain 2-IODO-*p*-XYLENE from *p*-xylene. The synthesis of a variety of symmetrically substituted biaryls is greatly improved by the reaction of Tl(I) bromide with appropriate aryl Grignard reagents; 4,4'-DIMETHYL-1,1'-BIPHENYL serves as a model. Lithium organocuprates continue to enjoy wide use: stereospecifically synthesized lithium vinylcuprates are utilized in the PREPARATION OF ALKENES BY REACTION OF LITHIUM DIPROPENYLCUPRATES WITH ALKYL HALIDES, and SECONDARY AND TERTIARY ALKYL KETONES are readily available FROM CARBOXYLIC ACID CHLORIDES AND LITHIUM PHENYLTHIO(ALKYL)CUPRATE REAGENTS. The ADDITION OF ORGANOLITHIUM REAGENTS TO ALLYL

ALCOHOL represents a convenient method of converting allyl alcohol to 2-substituted 1-propanols, while a one-pot reaction sequence of alkylation (alkyl lithium) and reduction (lithium–liquid ammonia) provides excellent yields of AROMATIC HYDRO-CARBONS FROM AROMATIC KETONES AND ALDEHYDES.

The application of phase-transfer catalysis is widespread and encompasses a variety of processes. Because of the operational simplicity and high yields of products, this technique will find great application in the future. Included here are PHASE-TRANSFER ALKYLATION OF NITRILES, using α-PHENYLBUTYRONIT-RILE as an example, and the HOFMANN CARBYLAMINE REACTION that involves the generation of dichlorocarbene. The preparation of 2-PHENYL-2-VINYLBUTYRONITRILE may be, in a sense, categorized in this class of reaction. A new, highly useful method to prepare ACETYLENIC ALDEHYDES AND KETONES *via* FRAGMENTATION OF α,β-EPOXYKETONES is described. The reagent required for this fragmentation is an *N*-AMINO-AZIRIDINE and a full description of experimental details for its preparation is given. The synthesis of enolizable β-dicarbonyl com-pounds by means of SULFIDE CONTRACTION *via* ALKYLATIVE COUPLING illustrates a general principle that two fragments can be combined by using the reactivity of sulfur which, in turn, is expelled from the condensate. Another fragmentation utilized for the stereo-selective synthesis of (*E*)-disubstituted alkenyl alcohols involves the reductive elimination of cyclic β-haloethers with sodium which, for example, converts 3-chloro-2-methyltetrahydropyran into (*E*)-4-HEXEN-1-OL. This product, after conversion into ethyl (*E*)-2-cyano-6-octenoate, is utilized to illustrate FREE RADICAL CYCLIZATION, initiated with dibenzoyl peroxide, and results in a high yield of ETHYL 1-CYANO-2-METHYLCYCLOHEXANE-CARBOXYLATE.

The recent upsurge of interest in systems of theoretical interest demands practical syntheses of several important compounds. These are BICYCLO[2.1.0]PENT-2-ENE, BENZOCYCLOPROPENE, 1,6-OXIDO[10]ANNULENE, and others. *endo*-TRICYCLO[4.4.0.0^{2,5}]-DECA-3,8-DIENE-7,10-DIONE is utilized as a model for the use of CYCLOBUTADIENE IN SYNTHESIS, and a stable monomeric ketene, *tert*-BUTYLCYANOKETENE offers opportunities for further studies of this interesting species.

Older reactions familiar to us have constantly been improved in the selectivity of reaction and yield of product. Thus three recent bromination techniques involve use of the following: (a) 2,4,4,6-tetrabromo2,5-cyclohexadien-1-one for *para*-BROMINATION OF AROMATIC AMINES, (b) the α-bromination of hexanoyl chloride with *N*-bromosuccinimide to provide 2-BROMOHEXANOYL CHLORIDE, and (c) use of methanol as a solvent to achieve monobromination at the less-substituted carbon atom of an unsymmetrical ketone as exemplified by the preparation of 1-BROMO-3-METHYL-2-BUTANONE. The well-known Lemieux–Rudloff oxidation is utilized to prepare 17β-HYDROXY-5-OXO-3,5-*seco*-4-NORANDROSTANE-3-CARBOXYLIC ACID, and OXIDATION WITH THE CHROMIUM TRIOXIDE-PYRIDINE COMPLEX PREPARED *in situ* simplifies the procedures utilized by its predecessors, the Sarett and Collins reagents. THE USE OF ION EXCHANGE RESIN FOR PREPARATION OF QUATERNARY AMMONIUM HYDROXIDES is preferred over conventional precipitation methods, and the overall process to prepare ALKENES *via* HOFMANN ELIMINATION has been improved. (—)-2,3:4,6-Di-*O*-isopropyliden-2-keto-*L*-gulonic acid hydrate [(—)-DAG] proves to be an efficient agent to resolve amines such as α-(1-NAPHTHYL)-ETHYLAMINE. Preparations of three useful compounds, METHYL NITROACETATE, 3-(4-CHLOROPHENYL)-5-(4-METHOXYPHENYL)-ISOXAZOLE, and 6,7-DIMETHOXY-3-ISOCHROMANONE also illustrate the most practical methods to date.

The Board of Editors is grateful to the contributors of the preparations included in this volume and welcomes both the submission of preparations for future volumes and suggestions for change that will enhance the usefulness of *Organic Syntheses*. The attention of submitters of preparations is drawn to the instructions on pages v and vi that describe the type of preparation we wish to obtain and also the information to be incorporated in each preparation. A style guide for preparing manuscripts is available from the Secretary to the Board, and the submitters are urged to follow its instructions.

The recent practice to list unchecked procedures at the end of each volume of *Organic Syntheses* has been slightly modified. Thus of the preparations received between July 1, 1974 and June 30, 1975, only those that have been accepted by the Board of Editors for checking appear as an insert in this volume. These unchecked procedures, available from the Secretary's office for a nominal fee, allow the experimental procedures we receive to be more rapidly accessible to interested users. The members of the Board will continue to check the accepted procedures before final publication.

The Editor-in-Chief is most grateful to Mr. Gordon S. Bates and Dr. P. E. Georghiou for their invaluable assistance in completing the final version of this volume, and to Dr. S. Kasparek for preparing the two sets of subject indexes arranged in both the conventional and systematic nomenclatures. Thanks are also due to Mrs. Laura Dodds and Miss Diane L. Dowhaniuk, who typed the entire manuscript.

Edmonton, Alberta, Canada   SATORU MASAMUNE
February 1975

CONTENTS

CONTENTS

ORGANIC SYNTHESES

ADDITION OF ORGANOLITHIUM REAGENTS TO ALLYL ALCOHOL: 2-METHYL-1-HEXANOL

$$\text{OH} + CH_3(CH_2)_3Li \xrightarrow[\text{pentane, 25°}]{(CH_3)_2NCH_2CH_2N(CH_3)_2}$$

Submitted by J. K. CRANDALL and A. C. ROJAS[1]
Checked by D. E. BERTHET and G. BÜCHI

1. Procedure

A 500-ml., three-necked, round-bottomed flask is fitted with a gas-inlet tube, a rubber septum, a reflux condenser connected to a mineral oil bubbler, and a sealed mechanical stirrer. The system is flamed with a Bunsen burner while flushing with dry nitrogen. The reaction vessel is cooled under nitrogen in an ice bath, and 7.25 g. (0.125 mole) of 2-propen-1-ol (Note 1), 70 ml. of pentane (Note 2), and 1.16 g. (0.010 mole) of N,N,N',N'-tetramethyl-1,2-ethanediamine (Note 3) are successively added through the rubber septum with a syringe. While maintaining a positive nitrogen pressure, 180 ml. of $1.5M$ (0.270 mole) butyllithium in pentane (Note 4) is added from a syringe over a 20-minute period (Note 5). The ice bath is removed and the reaction mixture is stirred for an additional hour (Note 6). The ice bath is then restored, the gas-inlet tube replaced with a pressure-equalizing dropping funnel, and 70 ml. of water is added, cautiously at first, and then more rapidly after the exothermic reaction ceases. The resulting mixture is transferred to a separatory funnel, the aqueous layer is separated and discarded, and the pentane layer is washed with a 10-ml. portion of aqueous $3N$ hydrochloric acid and then with two 10-ml. portions of water. The organic layer is dried over anhydrous magnesium sulfate, the drying agent is removed by filtration, and the solvent is removed by distillation through a 20-cm. Vigreux column.

1

Distillation of the residual oil through a short-path distillation apparatus yields 9.3–9.6 g. (64–66%) (Note 7) of 2-methyl-1-hexanol, b.p.
166–167° (Notes 8 and 9).

2. Notes

1. Commercial 2-propen-1-ol was purchased from Aldrich Chemical
Company, Inc. and was distilled prior to use (b.p. 94.5–95°).
2. Technical-grade pentane was distilled from concentrated sulfuric acid.
3. Commercial N,N,N',N'-tetramethyl-1,2-ethanediamine was obtained from Aldrich Chemical Company, Inc. and was distilled prior
to use (b.p. 119.5°).
4. Commercial solutions of butyllithium were obtained from Foote
Mineral Company.
5. During the addition of butyllithium, a gel forms. If the solution
is not well agitated during this period, the yield is somewhat lower.
6. Extending the reaction time did not increase the yield.
7. The checker's yield was 10.5–10.7 g. (72–74%) (see Note 5).
8. The purity of the product is greater than 99% as determined by
gas chromatographic analysis using a 6-m. column of 30% Carbowax
20M on 60–80 Chromosorb W. The major impurity ($<1\%$) was shown
to be 3-heptanol by comparison of gas chromatographic retention
times and mass spectral fragmentation patterns with those of an
authentic sample.
9. The spectral properties of the product are as follows; infrared
(neat) cm.$^{-1}$: 3268, 1377, 1037; proton magnetic resonance (carbon
tetrachloride) δ, multiplicity, number of protons: 0.88 (multiplet, 6),
1.38 (multiplet, 7), 3.33 (unresolved doublet, 2), 5.14 (broad singlet, 1).

3. Discussion

This procedure illustrates a convenient method of converting allyl
alcohol to 2-substituted-1-propanols by the addition of an organolithium reagent.[2] A variety of organolithiums have been demonstrated
to give moderate to high yields of the corresponding alcohols. The
indicated organolithium species is a demonstrated intermediate which
can, in principle, be employed in a host of further synthetic conversions.[2]
Substituted allylic alcohols, however, do not undergo analogous

conversions efficiently, except when the substituent is at the carbinol carbon.[2]

2-Methyl-1-hexanol has also been prepared by the reaction of 2-hexylmagnesium halides with formaldehyde,[3] the reduction of 2-methylhexanoic acid or its ester,[4,5] and by hydroformylation of 1-hexene[6-8] among others.

1. Department of Chemistry, Indiana University, Bloomington, Indiana 47401.
2. J. K. Crandall and A. C. Clark, *J. Org. Chem.*, **37**, 4236 (1972).
3. N. D. Zelinsky and E. S. Przewalsky, *J. Russ. Phys. Chem. Soc.*, **40**, 1105 (1908) [*C.A.*, **3**, 307 (1909)].
4. P. A. Levene and L. A. Mikeska, *J. Biol. Chem.*, **84**, 571 (1929).
5. M. Leclercq, J. Billard, and J. Jacques, *Mol. Cryst. Liquid Cryst.*, **8**, 367 (1969).
6. I. Wender, R. Levine, and M. Orchin, *J. Amer. Chem. Soc.*, **72**, 4375 (1950).
7. T. Asahara, H. Sekiguchi, and C. Kimura, *J. Soc. Org. Syn. Chem.* (*Tokyo*), **10**, 538 (1952) [*C.A.* **47**, 11123i (1953)].
8. British Petroleum Company Ltd., *Fr.* **1, 549, 414** (1968) [*C.A.*, **72**, 2995p (1970)].

ALKENES *via* HOFMANN ELIMINATION; USE OF ION-EXCHANGE RESIN FOR PREPARATION OF QUATERNARY AMMONIUM HYDROXIDES: DIPHENYLMETHYL VINYL ETHER

(Benzene, 1,1'-[(ethenyloxy)methylene]bis-)

$$(C_6H_5)_2CHOCH_2CH_2N(CH_3)_2 + CH_3\overset{-}{I} \xrightarrow{25°} (C_6H_5)_2CHOCH_2CH_2\overset{+}{N}(CH_3)_3\overset{-}{I}$$

$$\xrightarrow[\text{resin,methanol}]{\text{Anion-exchange}} [(C_6H_5)_2CHOCH_2CH_2\overset{+}{N}(CH_3)_3\overset{-}{OH}] \xrightarrow{\Delta} (C_6H_5)_2CHOCH=CH_2$$

Submitted by CARL KAISER and JOSEPH WEINSTOCK[1]
Checked by P. MÜLLER and G. BÜCHI

1. Procedure

A 250-ml., three-necked, round-bottomed flask equipped with a sealed mechanical stirrer, a dropping funnel, and a reflux condenser is charged with 13.3 g. (0.052 mole) of 2-(diphenylmethoxy)-*N,N*-dimethylethylamine (··· ethanamine) (Note 1) and 50 ml. of acetone. The solution is stirred, and 8.1 g. (0.057 mole) of iodomethane in 15 ml. of acetone is added dropwise over 5 minutes (Note 2). After the addition is complete, the mixture is stirred for 30 minutes, then cooled to 0–10° with an ice bath. The crystalline product is filtered and washed with 15 ml. of acetone and 30 ml. of ether to give 20.0–20.2 g. (97–98%) of colorless, crystalline methiodide, m.p. 194–196°.

An excess (60 g., *ca.* 0.26 equivalent) of anion exchange resin (OH⁻ form) (Note 3) in a 500-ml. Erlenmeyer flask is stirred with 200 ml. of methanol (Note 4) for 5 minutes. The methanolic slurry of resin is transferred to a 6.5 cm. by 25 cm. chromatography column, using 50–100 ml. of methanol to aid in the transfer. The resin column is washed with 750 ml. of methanol, added gradually so as to maintain about a 1–2.5 cm. solvent head above the upper resin level (Note 5). About two-thirds of the resin slurry is poured from the column (using about 100 ml. of methanol to aid the transfer) into a suspension of 19.9 g. (0.05 mole) of the methiodide in 50 ml. of methanol (Note 6). The mixture is stirred and heated gently on a water bath to dissolve the crystalline methiodide. The resulting resin suspension is poured onto the column containing the remaining one-third of the resin. Additional methanol (*ca.* 50 ml.) is required to facilitate transferral. The column is eluted with about 500 ml. of methanol until the eluent no longer affords an alkaline reaction to pH paper (Note 7). The methanolic eluent is concentrated under reduced pressure (10–25 mm.), and the residual liquid (Note 8) is gradually heated to 100° under the water-aspirator vacuum. Following completion of thermal decomposition, as evidenced by gas evolution (*ca.* 5–10 minutes), the residue is dissolved in 250 ml. of ether (Note 9). The ether solution is washed with 100 ml. of aqueous $0.2N$ sulfuric acid and 100 ml. of water, then dried over anhydrous magnesium sulfate, filtered, and the filtrate is concentrated. Distillation of the residue gives 8.5–9.0 g. (81–86%) of diphenylmethyl vinyl ether as a colorless liquid, b.p. 163–167° (18 mm.), n^{25} D 1.5716 (Notes 10 and 11).

2. Notes

1. The submitters used 2-(diphenylmethoxy)-N,N-dimethylethyl-amine, b.p. 150–165° (2 mm.).[2] This amine was obtained from Searle Chemicals, Inc. It is readily obtained from the hydrochloride, m.p. 161–162°, which is available commercially from Gane's Chemical Works, Inc., New York, N.Y., under the generic name, diphenhydramine.

2. The reaction exotherm is just sufficient to cause moderate reflux.

3. A strongly basic polystyrene alkyl quaternary amine (hydroxide form) of medium porosity was employed. Research-grade Rexyn

201 (OH) (purchased from Fisher Scientific Company) and Amberlite IRA-400 (purchased from Mallinckrodt Chemical Works) were found to be satisfactory. Chloride-form resins must be converted to the hydroxide form before use, as described below (Note 7).

4. It is necessary to wash the resin with methanol prior to packing of the column. If this is not done, swelling of the resin on treatment with the solvent may cause explosion of the column.

5. If the resin was not washed exhaustively with methanol, significant amounts of benzhydrol (α-phenyl-benzenemethanol) and diphenylmethyl methyl ether were obtained in the final product.

6. Stirring of the methiodide with the anion exchange resin prior to introduction into the column is necessary, because of the insolubility of this quaternary salt in methanol. For methanol-soluble methiodides, a solution of the salt may be added directly to the methanol-washed resin column.

7. The recovered resin can be reconverted to the hydroxide form by eluting a column of the material with aqueous 10% sodium hydroxide until it is free of halide ion (silver nitrate–nitric acid test) and then with water until the eluent is no longer alkaline to pH paper.

8. Heating should be carried out in a 1-l. (oversized) flask because decomposition is accompanied by considerable foaming.

9. A small amount of insoluble material, which is mainly unreacted 2-(diphenylmethoxy)-N,N-dimethylethylamine methiodide, can be removed at this point.

10. The product has the following spectral properties; infrared (neat) cm.$^{-1}$: 1670, 1200, 770, 710; proton magnetic resonance (chloroform-d) δ, multiplicity, number of protons, approximate coupling constants among the vinyl protons J in Hz.: 7.2 (singlet, 10), 6.36 (quartet, 1, $J_{AX} = 7$, $J_{AY} = 14$), 5.7 (singlet, 1), 4.27 (doublet of doublets, 1, $J_{XY} = 2$, $J_{AY} = 14$), 3.97 (doublet of doublets, 1, $J_{XY} = 2$, $J_{AX} = 7$). The distilled product was about 98% pure by gas chromatographic analysis on a 60 cm. by 0.6 cm. aluminum column packed with 10% SE 30 silicon rubber on Gas Chrom Z, 100–200 mesh, operated at 180°. The retention time was about 2.0 minutes. Minor amounts of benzhydrol and diphenylmethyl methyl ether (retention times 2.5 minutes and 1.8 minutes, respectively) accounted for the remainder of the distillate. The checkers found that gas chromatographic analysis on a 1.8-m. column packed with 15% SE 30 on GAW, 60–80 mesh at 180°, at an injector temperature of 250° resulted in

extensive decomposition. A satisfactory analysis, however, could be performed by lowering the injector temperature to 180°.

11. Reppe[3] reports b.p. 120° (15 mm.) for diphenylmethyl vinyl ether.

3. Discussion

Diphenylmethyl vinyl ether has also been prepared from benzhydrol and acetylene (ethyne) under high-pressure conditions.[3] In the described method, which is an adaptation of the procedure of Weinstock and Boekelheide,[4] improved yields of the alkene are obtained by using more convenient experimental conditions.

The described method for converting a quaternary halide to the corresponding hydroxide, utilizing an anion-exchange resin, has general application in the Hofmann elimination reaction.[5] It has been used extensively in the submitters' laboratories for the synthesis of a variety of alkenes[6] and for the preparation of a number of ethyl 1-benzylcyclopropanecarboxylates *via* abnormal Hofmann elimination of diethyl [2-(N,N-dimethylamino)ethyl]benzylmalonates.[7] It offers several notable advantages over more conventional methods for preparing quaternary hydroxides. Formation of quaternary hydroxides from iodides with bases (*e.g.*, silver oxide) that form insoluble iodides has disadvantages due to the expense of the reagent, and in some instances, from the oxidizing power of silver salts in basic solution. Thallous ethoxide has been used to avoid the oxidation effect; however, this is an expensive reagent[8-10] and is also toxic. Quaternary methosulfates may be hydrolyzed to sulfates and then converted to the hydroxide with barium hydroxide,[11] but this method has not found general application. The described procedure of exchange of hydroxide ion for halide is suitable for even very sensitive compounds and obviates most of the objectionable features of the precipitation methods.[4,5] In the event that methanol is undesirable, the conversion may be carried out in water.[4]

1. Smith Kline and French Laboratories, Philadelphia, Pennsylvania 19101.
2. G. Rieveschl, Jr., (to Parke Davis and Company), U.S. Patent **2, 421, 714** (1947) [*C.A.*, **41**, 5550h (1947)].
3. W. Reppe, *Justus Liebigs Ann. Chem.*, **601**, 81 (1956) [*C.A.*, **51**, 9579b (1957)].
4. J. Weinstock and V. Boekelheide, *J. Amer. Chem. Soc.*, **75**, 2546 (1953).
5. A. C. Cope, *Org. React.*, **11**, 317 (1960).
6. C. Kaiser and C. L. Zirkle (to Smith Kline and French Laboratories), U.S. Patent **3,462,491**, August 19, 1969.

7. C. Kaiser, C. A. Leonard, G. C. Heil, B. M. Lester, D. H. Tedeschi, and C. L. Zirkle, *J. Med. Chem.*, **13**, 820 (1970).
8. H. Wieland, C. Schöpf, and W. Hermsen, *Justus Liebigs Ann. Chem.*, **444**, 40 (1925).
9. B. Witkop, *J. Amer. Chem. Soc.*, **71**, 2559 (1949).
10. F. v. Bruchhausen, H. Oberembt, and A. Feldhaus, *Justus Liebigs Ann. Chem.*, **507**, 144 (1933).
11. J. v. Braun and E. Anton, *Ber.*, **64**, 2865 (1931).

AROMATIC HYDROCARBONS FROM AROMATIC KETONES AND ALDEHYDES: 1,1-DIPHENYLETHANE
(1,1'-Ethylidenebisbenzene)

$$C_6H_5COC_6H_5 \xrightarrow[\substack{\text{diethyl} \\ \text{ether, 25°}}]{CH_3Li} \left[(C_6H_5)_2\overset{\overset{O \ Li}{|}}{C}CH_3 \right] \xrightarrow[\substack{NH_4Cl, \ -33°}]{Li, \ NH_3 \ (liq.)} (C_6H_5)_2CHCH_3$$

Submitted by Sharon D. Lipsky and Stan S. Hall[1]
Checked by Robert E. Ireland, Paula J. Clendening,
Kathryn D. Crossland, and Alvin K. Willard

1. Procedure

A 500-ml., three-necked, round-bottomed flask, equipped with a 5-cm. glass-coated magnetic stirring bar, a Dewar condenser connected to a static argon line (Note 1), and a pressure-equalizing dropping funnel, is sealed with a rubber septum (Note 2). After flushing with argon, 30 ml. of anhydrous ether and 19.5 ml. of 1.89M (0.037 mole) methyllithium solution (Note 3) are injected through the septum into the flask (Note 4). A solution of 4.54 g. (0.025 mole) of benzophenone (diphenylmethanone) (Note 5) in 35 ml. of anhydrous ether is then placed in the dropping funnel and added to the reaction mixture over a 20-minute period. The mixture is allowed to stir for one half hour (Note 6). The septum is removed, and the side arm is quickly adapted with Tygon® tubing that leads through a tower of solid potassium hydroxide to a tank of anhydrous ammonia. After approximately 75 ml. of ammonia is slowly distilled into the flask (Note 7), the tubing is removed, and 0.525 g. (0.075 g.-atom, *ca.* 15 cm. added as 0.5 cm. pieces) of lithium wire (Note 8) is quickly added, and the flask is stoppered. After 15 minutes the dark blue color is discharged by the continuous addition of excess ammonium chloride (*ca.* 5 g. over a 15-minute period) (Note 9). The argon-inlet tube is disconnected, and the ammonia is allowed to evaporate. The residue is then partitioned between 100 ml. of aqueous saturated sodium chloride and 100 ml. of ether. The aqueous layer is separated and extracted with two 50-ml. portions of ether. The combined ether extracts are dried over anhydrous magnesium sulfate. Removal of the ether on a rotary evaporator yields 4.36–4.49 g. of crude product (Note 10). Filtration of the product through 60 g. of Woelm alumina (Grade III) with 150 ml. of petroleum ether affords 4.16–4.31 g. (92–95%) of 1,1-diphenylethane (Note 11), which on evaporative distillation in a Kugelrohr oven gives 4.12–4.21 g. (91–93%) of 1,1-diphenylethane, b.p. 100° (0.25 mm.), n^{28} D 1.5691 (Note 12).

2. Notes

1. The entire reaction sequence is performed under an argon atmosphere which is connected by a T-tube to the assembly and an oil bubbler, and is operated at a moderate flow-rate throughout the synthesis.

2. All of the glassware is oven dried and cooled to room temperature in a large box desiccator, or the assembled glassware can be flamed dry under an argon atmosphere and allowed to cool.

3. Methyllithium in ether solution is available from Foote Mineral Company.

4. If methyllithium of a different molarity is used, the total volume should be adjusted to 50 ml. by varying the amount of ether used.

5. Benzophenone is available from Matheson Coleman and Bell.

6. Toward the end of this sequence 2-propanol and dry ice are added to the condenser in preparation for the reduction step.

7. To prevent splattering, the apparatus is tilted slightly to allow the condensing ammonia to run down the walls of the flask.

8. Evidently surface area is important, since when the 15 cm. of lithium wire was added as 1-cm. pieces, the reduction was incomplete. Lithium wire (0.32 cm., 0.01% sodium) available from Alpha Inorganics, Inc. was wiped free of oil and rinsed with petroleum ether immediately prior to use.

9. The ammonium chloride is most conveniently introduced by attaching a glass tube filled with the salt to a side arm by means of Tygon® tubing. When the ammonium chloride is to be added, the tube is raised and tapped gently to introduce the quenching agent smoothly. Often a very vigorous reaction occurs after a considerable (10-minute) induction period. Should this step start to become violent,

the addition and the vigorous stirring should be momentarily stopped to avoid eruption.

10. Gas chromatography on a 200 cm. by 0.6 cm. column packed with 10% Apiezon L on Chromosorb W (AW, DMCS) using a flame-detector instrument, at a 40 ml./minute helium carrier gas flow rate, gives a trace peak at 9.9 minutes (diphenylmethane), a major peak at 11.7 minutes (1,1-diphenylethane), and a trace peak at 15.4 minutes (1,1-diphenylethanol) when the oven is held at 190° for 10 minutes and then programmed at 10°/minute to 290°.

11. The sample is sufficiently pure at this point to use for most purposes. The chromatography step is an efficient means to remove any 1,1-diphenylethanol that was not reduced.

12. The spectral properties of the product are as follows; proton magnetic resonance (carbon tetrachloride) δ, multiplicity, number of protons, assignment, coupling constant J in Hz.: 1.54 (doublet, 3, CH_3, $J = 7$), 4.03 (quartet, 1, CH, $J = 7$), 7.12 (singlet, 10, 2 $\times$ C_6H_5); mass spectrum m/e (relative intensity): 182 (M, 32), 167 (100).

3. Discussion

This procedure illustrates a general method for preparing aromatic hydrocarbons by the tandem alkylation–reduction of aromatic ketones and aldehydes.[2] Additional examples are given in Table I.

The advantages of the method are that the entire sequence is carried out in the same reaction vessel without isolation or purification of intermediates. The procedure consumes only a few hours, and in most cases the isolated yield of the aromatic hydrocarbon is excellent.

The method may be modified so that the organolithium reagent is generated in situ in ether from the corresponding bromide. Best results were obtained by having all of the lithium wire necessary to generate the organolithium reagent and to reduce the intermediate benzyl alkoxide present from the outset.[3] Commercial organolithium reagents such as butyllithium in hexane or phenyllithium in ether–benzene were satisfactory when twice as much lithium is used for the reduction step. In some cases, by running the alkylation step at −78° to minimize competing side reactions,[4] higher yields than those listed in Table I can be realized.

In addition to the present method, other procedures have been reported for the synthesis of 1,1-diphenylethane.[5-7]

TABLE I

Aromatic Hydrocarbons from Aromatic Ketones and Aldehydes

Aromatic Carbonyl Compound	Organolithium Reagent	Product	Yield(%)
	CH_3Li		95
	CH_3Li		94
	CH_3Li		95
	CH_3Li		78
	C_4H_9Li	C_5H_{11}	86[a,e]
	C_4H_9Li	C_4H_9	89[a,e]
	C_4H_9Li	C_4H_9	76[b,e]
	C_4H_9Li	C_4H_9	70[a,e]

10

TABLE I (*Cont.*)

Aromatic Carbonyl Compound	Organolithium Reagent	Product	Yield(%)
	C_6H_5Li		93[c,e]
	C_6H_5Li		89[d,e]
	C_6H_5Li		97[c,e]
	C_6H_5Li		97[d,e]

[a] The organolithium reagent was generated *in situ* in ether from 1-bromobutane.
[b] Commercial butyllithium (Foote Mineral Company, *ca.* 15% in hexane) and six equivalents of lithium were used.
[c] The organolithium reagent was generated *in situ* in ether from bromobenzene.
[d] Commercial phenyllithium (Foote Mineral Company, *ca.* 20% in ether–benzene) and six equivalents of lithium were used.
[e] Reaction conducted on a 0.005 mole scale using as solvent 20 ml. of ether and 20 ml. of ammonia. Yield after column chromatography.

1. Department of Chemistry, Rutgers University, Newark, New Jersey 07102.
2. S. S. Hall and S. D. Lipsky, *Chem. Commun.*, 1242 (1971); *J. Org. Chem.*, **38**, 1735 (1973).
3. The lithium wire is cut into 0.5-cm. pieces and hammered to a foil immediately prior to use.
4. J. D. Buhler, *J. Org. Chem.*, **38**, 904 (1973).
5. J. S. Reichert and J. A. Nieuwland, *J. Amer. Chem. Soc.*, **45**, 3090 (1923).
6. E. Späth, *Monatsh. Chem.*, **34**, 1965 (1913).
7. J. Böeseken and M. C. Bastet, *Rec. Trav. Chim. Pays-Bas*, **32**, 184 (1913).

BENZOCYCLOPROPENE

(Bicyclo [4.1.0] hepta-1,3,5-triene)

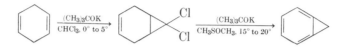

Submitted by W. E. Billups, A. J. Blakeney, and W. Y. Chow[1]
Checked by Nobuo Nakamura and S. Masamune

1. Procedure

Caution! Benzocyclopropene is characterized by an extremely unpleasant (foul) odor, and use of a good hood is recommended for the preparation.

A. *7,7-Dichlorobicyclo[4.1.0]hept-3-ene.*[2] A 2-l., three-necked, round-bottomed flask is equipped with a sealed mechanical stirrer, a reflux condenser, and a pressure-equalizing dropping funnel. The system is flushed with nitrogen by means of a gas-inlet tube attached to the top of the condenser, and 126 g. (1.123 mole) of potassium *tert*-butoxide (Note 1) and 1.2 l. of pentane are added. The stirred suspension is cooled to 0–5° with an ice bath and 90 g. (1.123 mole) of 1,4-cyclohexadiene (Note 2) is introduced rapidly through the dropping funnel, and 135 g. (1.131 mole) of chloroform (Note 3) is then added dropwise over a period of 1.5–2 hours. The resulting mixture is stirred for an additional 30 minutes, and 300 ml. of cold water is then added to dissolve all of the precipitated salts. The organic phase is separated, and the aqueous phase is extracted once with a 50-ml. portion of pentane. The extract is combined with the original pentane solution and dried over approximately 20 g. of anhydrous sodium sulfate. The solvent is then removed on a rotary evaporator, and the product is distilled through a 15-cm. Vigreux column to give 69–72 g. (38–39%) of 7,7-dichlorobicyclo-[4.1.0]hept-3-ene, b.p. 50–51° (0.8 mm.) (Note 4).

B. *Benzocyclopropene.* A dry, three-necked, round-bottomed flask fitted with a sealed mechanical stirrer, a reflux condenser, and a pressure-equalizing dropping funnel is flushed with nitrogen. To the flask is added 35.0 g (0.312 mole) of potassium *tert*-butoxide (Note 1), followed by 200 ml. of dimethyl sulfoxide (Note 5). The stirred mixture is cooled to 15–20° (Note 6) with an ice bath and 24.5 g.

(0.15 mole) of the 7,7-dichlorobicyclo[4.1.0]hept-3-ene is added over a 7-minute period. The bath is removed, the mixture stirred an additional 25 minutes, and the reaction quenched by first cooling the flask with an ice bath and then adding 180 ml. of ice water. The crude product is then pumped directly into an acetone–dry ice cold trap through a glass vacuum take-off adapter. The *tert*-butyl alcohol and dimethyl sulfoxide are removed by washing the distillate once with 400 ml. of ice water (Notes 7 and 8). The benzocyclopropene that separates as the lower layer is distilled from 1 g. of anhydrous sodium sulfate, using the apparatus shown in Figure 1. This procedure gives 4.35–5.48 g. (32–41%) of almost pure benzocyclopropene (Note 9).

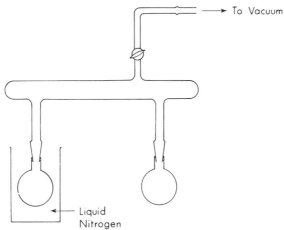

Figure 1.

2. Notes

1. Potassium *tert*-butoxide from a freshly opened bottle supplied by the MSA Research Corporation was used.

2. The checkers purchased 1,4-cyclohexadiene from Aldrich Chemical Company, Inc. and distilled it prior to use.

3. Reagent-grade chloroform was used without removal of stabilizer.

4. The product was shown to be approximately 95% pure by gas chromatographic analysis, using a 180 cm. by 0.24 cm. column packed with UCW-98 and heated to 130°.

5. Dimethyl sulfoxide (supplied by Aldrich Chemical Company, Inc.) was dried by distilling over calcium hydride at 5 mm.

6. Care should be taken not to freeze the dimethyl sulfoxide.

7. The checkers observed on one occasion that the mixture formed an emulsion and that centrifugation facilitated the separation of two layers.

8. An alternative procedure is to extract the product into pentane.

9. The proton magnetic resonance spectrum shows that the product is approximately 95% pure, the major impurities being toluene and styrene. The product has the following spectral properties; infrared (neat) cm.$^{-1}$: 1666, 1380, 1060, 735; proton magnetic resonance (chloroform-d) δ, number of protons: 3.17 (2), 7.21 (4); carbon-13 magnetic resonance (chloroform-d, tetramethylsilane reference) δ: 18.56, 115.07, 125.86, 129.15; ultraviolet (cyclohexane) nm. max. (log ϵ): 263 (3.88), 268 (3.96), 276 (3.90).

3. Discussion

Benzocyclopropene is an intriguing example in which the electronic structure of benzene is greatly perturbed by the fusion of the smallest alicyclic ring, cyclopropene, to the aromatic system. Benzocyclopropene thus arouses theoretical interest and the high strain energy (approximately 68 kcal./mole)[3] associated with the compound suggests unusual chemical reactivity. A review article has recently appeared.[4]

The first successful synthesis of benzocyclopropene was reported by Vogel and coworkers[5] and is illustrated below. Though elegant, this method does require the prior, lengthy synthesis of the commercially unavailable 1,6-methano[10]annulene.[6]

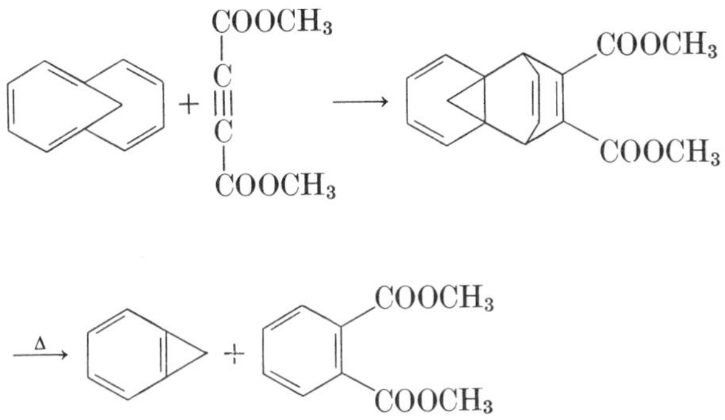

The procedure described here[7] is a convenient two-step reaction that relies on the base-induced elimination–isomerization reactions of gem-dichlorocyclopropanes.[8–15] The reaction mechanism has been studied.[16] The principal advantage of this method is the ready availability of necessary reagents.

1. Department of Chemistry, Rice University, Houston, Texas, 77001.
2. B. S. Farah and E. E. Gilbert, *J. Chem. Eng. Data*, **7**, 568 (1962).
3. W. E. Billups, W. Y. Chow, K. H. Leavel, E. S. Lewis, J. L. Margrave, R. L. Sass, J. J. Shieh, P. J. Werness, and J. L. Wood, *J. Amer. Chem. Soc.*, **95**, 7878 (1973).
4. B. Halton, *Chem. Rev.*, **73**, 113 (1973).
5. E. Vogel, W. Grimme, and S. Korte, *Tetrahedron Lett.*, 3625 (1965).
6. E. Vogel and H. D. Roth, *Angew. Chem.*, **76**, 145 (1964); *Angew. Chem. Int. Ed. Engl.*, **3**, 228 (1964); E. Vogel, W. Klug, and A. Breuer, *Org. Syn.*, **54**, 11 (1974).
7. W. E. Billups, A. J. Blakeney, and W. Y. Chow, *Chem. Commun.*, 1461 (1971).
8. C. L. Osborn, T. C. Shields, B. A. Shoulders, J. F. Krause, H. V. Cortez, and P. D. Gardner, *J. Amer. Chem. Soc.*, **87**, 3158 (1965).
9. T. C. Shields, B. A. Shoulders, J. F. Krause, C. L. Osborn, and P. D. Gardner, *J. Amer. Chem. Soc.*, **87**, 3026 (1965).
10. T. C. Shields and P. D. Gardner, *J. Amer. Chem. Soc.*, **89**, 5425 (1967).
11. T. C. Shields and W. E. Billups, *Chem. Ind. London*, 1999 (1967).
12. T. C. Shields, W. E. Billups, and A. R. Lepley, *J. Amer. Chem. Soc.*, **90**, 4749 (1968).
13. T. C. Shields and W. E. Billups, *Chem. Ind. London*, 619 (1969).
14. W. E. Billups, K. H. Leavell, W. Y. Chow, and E. S. Lewis, *J. Amer. Chem. Soc.*, **94**, 1770 (1972).
15. W. E. Billups, T. C. Shields, W. Y. Chow, and N. C. Deno, *J. Org. Chem.*, **37**, 3676 (1972).
16. J. Prestien and H. Günther, *Angew. Chem.*, **86**, 278 (1974); *Angew. Chem. Int. Ed. Engl.*, **13**, 276 (1974).

BICYCLO[2.1.0]PENT-2-ENE

Submitted by A. Harry Andrist, John E. Baldwin, and Robert K. Pinschmidt, Jr.[1]
Checked by Keith C. Bishop, III and Kenneth B. Wiberg

1. Procedure

Caution! The photochemical reaction should be carried out under a light absorbent cover. The operator should wear goggles affording protection from ultraviolet light.

While still slightly warm from the drying oven, the photolysis vessel with a water-jacketed quartz immersion well (Note 1) (section *A* of Figure 2) is charged with 500 ml. of anhydrous tetrahydrofuran (Note 2) and 10 ml. (8.05 g., 0.122 mole) of cyclopentadiene (Note 3). The solution is cooled in an ice bath and purged with dry nitrogen for 2 minutes. Then the vessel is sealed, the lamp inserted, and the solution irradiated at 0° for 30 minutes. During this period, sections *B* and *C* (see diagram) are removed from the oven to cool slightly. While still warm, the bottom part of C is filled with 0.5 ml. of anhydrous tetrahydrofuran and approximately 25 μl. of triethylamine (Note 4), greased, and stoppered. The top parts of B and C are thoroughly greased and connected to the 24/40 $\overline{\textbf{S}}$ joint of A. The bottom of part B is filled rapidly with 200 ml. of an approximately $1M$ solution of dimsyl anion (Note 5) in dimethyl sulfoxide. The bottom part of C is then connected and cooled in a 2-propanol–dry ice bath. The system is irradiated and subjected to a steady flow of nitrogen (approximately 400–500 ml./minute). After the first hour, this flow rate is reduced to 200–250 ml./minute for the remaining 4.5 hours. At the end of a total irradiation period of 6 hours, trap C is stoppered and stored at dry ice temperature. This trap contains 15–25 ml. of a 2–5% solution of pure bicyclopentene in tetrahydrofuran (Note 6), which can be stored indefinitely at $-78°$ (Notes 7 and 8).

2. Notes

1. The submitters used an unfiltered Hanovia 450-watt medium-pressure mercury vapor lamp (catalog No. 679A, Hanovia Lamp Division, Canrad Precision Industries, 100 Chestnut Street, Newark, N.J. 07105) and a reactor equipped with a nitrogen inlet containing a glass frit which entered the bottom of the reactor. All connections (Note 9) were glass with the exception of the rubber tubing between the second trap and the mercury bubbler. All glassware was soaked for 30 minutes in aqueous 30% ammonium hydroxide and oven-dried prior to use without further rinsing. Any polymer formed on the quartz immersion well during this preparation may be removed with nitric acid or with warm water and cleansing powder; if it is not removed, yields in subsequent preparations are significantly diminished.

2. Tetrahydrofuran (purchased from Mallinckrodt Chemical Works) was dried by distillation from lithium aluminum hydride, and stored over Linde 4A Molecular Sieves.

3. Cyclopentadiene was prepared[2] by heating dicyclopentadiene (purchased from Eastman Organic Chemicals) and a pinch of hydroquinone (1,4-benzenediol) under a column of glass helices or a Vigreux column at 175° and collecting the distillate in a receiver cooled with a 2-propanol–dry ice bath. The monomer was dried over Linde 4A Molecular Sieves at $-20°$ and could be stored at this temperature for several weeks without excessive dimerization.

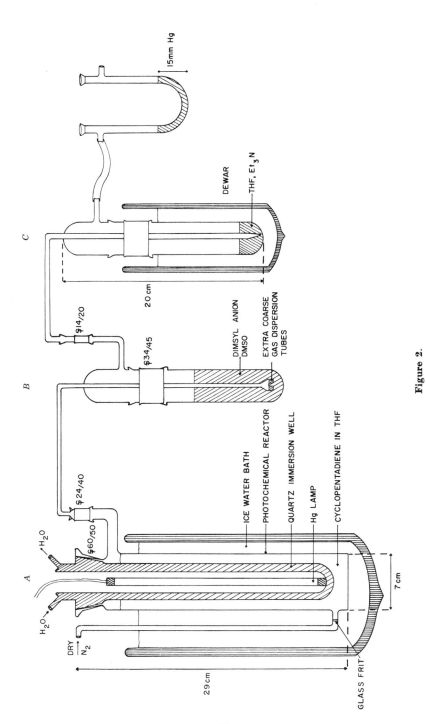

Figure 2.

15mm Hg

DEWAR

THF, Et$_3$N

C

20 cm

$\overline{\underline{S}}$14/20

$\overline{\underline{S}}$34/45

DIMSYL ANION
DMSO

EXTRA COARSE
GAS DISPERSION
TUBES

B

$\overline{\underline{S}}$24/40

$\overline{\underline{S}}$60/50

H$_2$O

A

H$_2$O

DRY
N$_2$

ICE WATER BATH

PHOTOCHEMICAL REACTOR

QUARTZ IMMERSION WELL

Hg LAMP

CYCLOPENTADIENE IN THF

7 cm

GLASS FRIT

29 cm

17

4. Triethylamine (purchased from Matheson Coleman and Bell) was dried over Linde 4A Molecular Sieves.

5. Dimsyl anion[3] was prepared from 10.2 g. (0.24 mole) (Note 10) of 56.8% sodium hydride, which was washed with pentane and vacuum dried, and 200 ml. of anhydrous dimethyl sulfoxide. The mixture was heated at 65–70° for about 50 minutes, until hydrogen evolution ceased. *Caution! This mixture should not be heated above 80°, because of the possibility of explosive decomposition.*

6. Analysis and purification of the product solution is best accomplished by gas chromatography. The submitters used a 500 cm. by 0.6 cm. aluminum or polyethylene column packed with 21% β,β'-oxydipropionitrile on Chromosorb P with column, injector and detector operated at 25° and a flow rate of 50 ml./minute. Under these conditions the retention times of bicyclopentene and cyclopentadiene were 3 and 5 minutes, respectively, beyond that of the coinjected air. Since bicyclopentene is extremely labile with respect to acid catalysis any contact with water, hydroxylic solvents, and nonprotic acids should be avoided (Note 11). Bicyclopentene stored at −78° in anhydrous tetrahydrofuran is stable indefinitely.

7. Decalin (decahydronaphthalene), benzene, 1,4-dioxane, and ethanol may be used as solvents for the photolysis. In an alternative procedure, volatile materials swept from the photolysis vessel are condensed in a dry ice trap. This cold mixture is added to a flask containing a magnetically stirred solution of dimsyl anion in dimethyl sulfoxide, and fractionation at reduced pressure provides a solution of bicyclopentene in tetrahydrofuran.

8. A purified, undiluted sample of bicyclopentene has been reported to explode.[4]

9. All of the connections must be well secured by sturdy rubber bands to avoid leakage caused by the *not* insubstantial back pressure that develops in the course of the reaction.

10. A higher base concentration or substitution of one or two gas-dispersion tubes leads to clogging of the inlet to the trap.

11. Two injections of 10 ml. of ammonia vapors greatly helps to eliminate decomposition on the column during gas chromatographic collection.

3. Discussion

The present procedure, a modification of that previously reported,[5–7] permits the ready preparation of a cyclopentadiene-free solution of bicyclopentene on a synthetic scale. Until now, synthetic chemistry involving the use of bicyclopentene has been limited to preparations of bicyclopentane[5,8] and bicyclopentane-*2,3-d*$_2$,[8] which is more a reflection of the extraordinary care required in handling this unusually sensitive bicyclic alkene than a lack of interesting potential or useful reactivity.

Most cyclic and acyclic 1,3-dienes, such as cyclopentadiene, undergo photochemical ring-closure to cyclobutenes.[9] Cyclopentadiene-*5-d*, cyclopentadiene-*d*$_6$, 2-methylcyclopentadiene,[10] and 1-methylcyclopentadiene[11] have been converted to the corresponding bicyclo-[2.1.0]pent-2-ene derivatives.

1. Department of Chemistry, University of Oregon, Eugene, Oregon 97403.
2. G. Wilkinson, *Org. Syn.*, Coll. Vol. 4, 473 (1963), cf. Note 4. See also M. Korach, D. R. Nielson, and W. H. Rideout, *Org. Syn.*, Coll. Vol. 5, 414 (1973).
3. E. J. Corey and M. Chaykovsky, *J. Amer. Chem. Soc.*, 84, 866 (1962); E. J. Corey and M. Chaykovsky, *J. Amer. Chem. Soc.*, 87, 1345 (1965).
4. T. J. Katz, private communication.
5. J. I. Brauman, L. E. Ellis, and E. E. van Tamelen, *J. Amer. Chem. Soc.*, 88, 846 (1966).
6. J. I. Brauman and D. M. Golden, *J. Amer. Chem. Soc.*, 90, 1920 (1968).
7. D. M. Golden and J. I. Brauman, *Trans. Faraday Soc.*, 65, 464 (1969).
8. P. G. Gassman, K. T. Mansfield, and T. J. Murphy, *J. Amer. Chem. Soc.*, 91, 1684 (1969); P. G. Gassman, K. T. Mansfield, and T. J. Murphy, *J. Amer. Chem. Soc.*, 90, 4746 (1968).
9. G. B. Gill, *Quart. Rev. Chem. Soc.*, 22, 338 (1968); G. J. Fonken, *Org. Photochem.*, 1, 197 (1967).
10. J. E. Baldwin and A. H. Andrist, *Chem. Commun.*, 1561 (1970).
11. J. E. Baldwin and G. D. Andrews, *J. Amer. Chem. Soc.*, 94, 1775 (1972).

para-BROMINATION OF AROMATIC AMINES:
4-BROMO-*N*,*N*-DIMETHYL-3-(TRIFLUOROMETHYL)ANILINE

[Benzenamine, 4-bromo-*N*,*N*-dimethyl-3-(trifluoromethyl)]

Submitted by G. J. Fox,[1] G. Hallas,[2]
J. D. Hepworth,[1] and K. N. Paskins[2]
Checked by J. D. Lock, Jr. and S. Masamune

1. Procedure

Caution! The reaction should be conducted in a hood to avoid inhalation of bromine vapor.

A. *2,4,6-Tetrabromo-2,5-cyclohexadien-1-one.* A mixture of 66.2 g. (0.2 mole) of 2,4,6-tribromophenol (Note 1), 27.2 g. (0.2 mole) of sodium acetate trihydrate, and 400 ml. of glacial acetic acid is placed in a 1-l. Erlenmeyer flask and warmed until a clear solution is obtained. The temperature of the solution is approximately 70°. The solution is magnetically stirred and cooled to room temperature to produce a finely divided suspension of the phenol. A solution of 32 g. (0.2 mole) of bromine in 200 ml. of glacial acetic acid is added dropwise over 1 hour (Note 2). The resulting mixture is kept at room temperature for 30 minutes and is then poured onto 2 kg. of crushed ice. The yellow solid which separates is removed by suction filtration after the ice has melted, and the damp crystals are dissolved in the minimum quantity of warm chloroform (Note 3). The upper aqueous layer is removed by means of a pipet fitted with a suction bulb. The dienone crystallizes from the

20

chloroform solution upon cooling, yielding 50–55 g. (61–67%) of crystals, m.p. 125–130° (decomp.), which are of sufficient purity for use in the next step (Notes 4 and 5).

B. *4-Bromo-N,N-dimethyl-3-(trifluoromethyl)aniline.* A solution of 9.45 g. (0.05 mole) of *N,N*-dimethyl-3-(trifluoromethyl)aniline (Note 6) in 200 ml. of dichloromethane is placed in a 500-ml. Erlenmeyer flask and cooled to −10°. To the magnetically stirred solution is added 20.5 g. (0.05 mole) of finely powdered 2,4,4,6-tetrabromo-2,5-cyclohexadien-1-one in approximately 0.5-g. portions. During this addition the temperature of the mixture should be maintained between −10° and 0° (Note 7). The cooling bath is removed, the reaction mixture is allowed to warm to room temperature over a 30-minute period and is extracted twice with 50 ml. of aqueous 2*N* sodium hydroxide in order to remove 2,4,6-tribromophenol (Note 8). The organic layer is washed with 25 ml. of water and dried over anhydrous magnesium sulphate. Removal of the solvent affords 12–12.5 g. of crude 4-bromo-*N,N*-dimethyl-3-(trifluoromethyl)aniline. Distillation through an 8-cm. Vigreux column provides 11–12 g. (82–90%) of pure bromoamine, b.p. 134–136° (15 mm.), which solidifies to give colorless crystals, m.p. 29–30° (Notes 9 and 10).

2. Notes

1. The submitters used reagent-grade 2,4,6-tribromophenol. The checkers recrystallized the practical-grade reagent purchased from Fisher Scientific Company. The solvent used was Skelly B, and the melting point of the phenol, after recrystallization, was 93–95°.

2. It is essential to maintain the temperature of the solution below 25° during the addition of the bromine solution. If external cooling is applied, initially with an ice-water bath, the addition can be completed within 20 minutes.

3. Some decomposition of the dienone is observed if the chloroform solution is vigorously refluxed for any length of time; bromine is evolved and there is a reduction in yield. Approximately 400 ml. of chloroform is needed to dissolve the dienone at approximately 60°. The checkers used 450 ml. of the solvent.

4. The submitters used 0.5-molar quantities with no reduction in yield.

5. Proton magnetic resonance (dioxane-d_8) δ, multiplicity: 7.98 (singlet).

6. The amine was prepared according to the procedure described by W. A. Sheppard in *Org. Syn.*, Coll. Vol. **5**, 1085 (1973).

7. The reaction proceeds satisfactorily over a range between $-30°$ and $+20°$. At lower temperatures, the reaction proceeds rather slowly.

8. 2,4,6-Tribromophenol may be recovered by acidification of the aqueous alkaline extracts and can be reapplied in the preparation of the tetrabromo-compound after crystallization from petroleum ether (b.p. 80–100°).

9. The product can be crystallized from petroleum ether (b.p. 30–40°).

10. Proton magnetic resonance (chloroform-*d*) δ, multiplicity, number of protons, assignment: 2.94 (singlet, 6, 2 $\times$ CH_3), 6.7 (approximate doublet of doublets, 1), 7.0 (approximate doublet, 1), 7.5 (approximate doublet, 1).

3. Discussion

4-Bromo-*N*,*N*-dimethyl-3-(trifluoromethyl)aniline has been prepared by the methylation of 4-bromo-3-(trifluoromethyl)aniline with trimethyl phosphate in 70–80% yield.[3] The present method, which effectively uses 3-(trifluoromethyl)aniline as starting material, offers advantages in cost, yield, and ease of purification.

Aromatic amines are usually polybrominated on treatment with bromine. Several mild brominating agents have been introduced in attempts to achieve partial bromination without the necessity of protecting the amino group and subsequent hydrolysis, but these give variable results when applied to a large variety of amines. Dioxane dibromide[4] monobrominates tertiary aromatic amines, but gives poor yields with primary and secondary aryl amines. The use of *N*-bromosuccinimide[5,6] (1-bromo-2,5-pyrrolidinedione) leads to monobrominated compounds that are frequently contaminated with decomposition products.

The dienone, which is prepared essentially as described by Benedikt[7] and Caló,[8] monobrominates a wide range of primary, secondary, and tertiary aromatic amines almost exclusively in the *para*-position. The procedure described is of general synthetic utility for the preparation of *para*-brominated aromatic and heteroaromatic amines in high yields and frequently in a high state of purity. The submitters have used this technique to *para*-brominate many compounds in quantities ranging within 0.01–0.1 mole, including the following in the yields, after one crystallization, indicated: aniline (benzenamine) (92), *N*-methylaniline (94), *N,N*-dimethylaniline (91), *N,N*-diethylaniline (94), *o*-toluidine (2-methylbenzenamine) (88), 2-methyl-*N*-methylaniline (91), 2-methyl-*N,N*-dimethylaniline (90), *m*-toluidine (90), 3-methyl-*N,N*-dimethylaniline (86), 2,3-dimethylaniline (91), 2,5-dimethylaniline (91), 3,5-dimethylaniline (81), 2-chloroaniline (86), 3-chloroaniline (86), 2-bromoaniline (78), 3-bromoaniline (82), 2-nitroaniline (91), 3-nitroaniline (85), *m*-phenylenediamine (82), *o*-anisidine (2-methoxy-benzenamine) (85), 3-methoxyaniline (58, 4-bromo; 30, 6-bromo), diphenylamine (90), 1-aminonaphthalene (1-naphthalenamine) (86), 1-dimethylaminonaphthalene (84), 2-(trifluoromethyl)aniline (85), 2-aminopyridine (2-pyridinamine) (75), 2-dimethylaminopyridine (70), 3-dimethylaminopyridine (60),[9] 2-amino-6-methylpyridine (76). Where solubility of the amine in dichloromethane is low, chloroform can be used as solvent. For example, 2-aminopyrimidine gave 2-amino-5-bromopyrimidine (82%) in this manner, compared with 41% when the amine is brominated conventionally in aqueous solution.[10] In the case of anthranilic acid (2-aminobenzoic acid), 5-bromo-2-aminobenzoic acid (82%) precipitated from the chloroform reaction medium. In addition to its use with amines, the dienone reagent monobrominates a variety of phenols,[11] and behaves as an oxidizing agent toward sulfides, converting them to sulfoxides.[12]

1. Department of Chemistry, College of Art and Technology, Derby, England. Present address: Department of Chemistry and Biology, Preston Polytechnic, Preston, England.
2. Department of Colour Chemistry, The University, Leeds, England.
3. D. E. Grocock, G. Hallas, and J. D. Hepworth, *J. Chem. Soc. Perkin II*, 1792 (1973).
4. G. M. Kosolapoff, *J. Amer. Chem. Soc.*, **75**, 3596 (1953).
5. L. Horner, E. Winkelmann, K. H. Knapp, and W. Ludwig, *Chem. Ber.*, **92**, 288 (1959).
6. J. B. Wommack, T. G. Barbee, Jr., D. J. Thoennes, M. A. McDonald, and D. E. Pearson, *J. Heterocycl. Chem.*, **6**, 243 (1969).
7. R. Benedikt, *Justus Liebigs Ann. Chem.*, **199**, 127 (1879).
8. V. Caló, F. Ciminale, L. Lopez, and P. E. Todesco, *J. Chem. Soc. C*, 3652 (1971).
9. G. J. Fox, J. D. Hepworth, and G. Hallas, *J. Chem. Soc. Perkin I*, 68 (1973).
10. J. P. English, J. H. Clark, J. W. Clapp, D. Seeger, and R. H. Ebel, *J. Amer. Chem. Soc.*, **68**, 453 (1946).
11. V. Caló, F. Ciminale, L. Lopez, G. Pesce, and P. E. Todesco, *Chimica e Industria*, **53**, 467 (1971).
12. V. Caló, F. Ciminale, G. Lopez, and P. E. Todesco, *Int. J. Sulphur Chem.*, Part A, **1**, 130 (1971).

1-BROMO-3-METHYL-2-BUTANONE

$$(CH_3)_2CH-\underset{\underset{O}{\|}}{C}-CH_3 + Br_2 \xrightarrow[0° \text{ to } 5°]{CH_3OH} (CH_3)_2CH-\underset{\underset{O}{\|}}{C}-CH_2Br + HBr$$

Submitted by M. Gaudry and A. Marquet[1]
Checked by Diana Metzger and Richard E. Benson

1. Procedure

Caution! This preparation must be carried out in an efficient hood. Bromomethyl ketones are highly lachrymatory and are skin irritants.

A 2-l., four-necked, round-bottomed flask equipped with a sealed mechanical stirrer, a thermometer, a reflux condenser fitted with a calcium chloride drying tube, and a 100-ml. pressure-equalizing dropping funnel is charged with 86 g. (105 ml., 1.0 mole) of 3-methyl-2-butanone (Note 1) and 600 ml. of anhydrous methanol (Note 2). The solution is stirred and cooled in an ice-salt bath to 0–5°, and 160 g. (54.6 ml., 1.0 mole) of bromine (Note 3) is added in a rapid, steady stream from the dropping funnel (Note 4). During this time, the temperature is allowed to rise but is not permitted to exceed 10°. The reaction temperature is then maintained at 10° during the remaining reaction time (Note 5). The red color of the solution fades gradually in about 45 minutes (Note 6), 300 ml. of water is then added (Note 7), and the mixture is stirred at room temperature overnight (Note 8).

To the solution is added 900 ml. of water, and the resulting mixture is washed with four 500-ml. portions of ether. The ether layers are combined and washed with 200 ml. of aqueous 10% potassium carbonate and then twice with 200-ml. portions of water (Note 9). The ether layer is dried for 1 hour over 200 g. of anhydrous calcium chloride (Note 10) and the solvent is removed on a rotary evaporator at room temperature to give 145–158 g. of crude product (Note 11). Distillation under reduced pressure through a Vigreux column gives 115–128 g. of a fraction, b.p. 83–86° (54 mm.), n^{22} D 1.4620–1.4640, containing 95% of 1-bromo-3-methyl-2-butanone as established by proton magnetic resonance measurements (Note 11).

2. Notes

1. The checkers used 3-methyl-2-butanone purchased from Eastman Organic Chemicals. One sample that gave a positive test for peroxides was purified by passage through a column of alumina before distillation. The material was distilled routinely before use.

2. The methanol was distilled twice from magnesium turnings. Alternately, it was dried overnight over molecular sieves and then distilled. The checkers also found freshly opened reagent-grade methanol (purchased from Fisher Scientific Company) to be satisfactory.

3. The submitters used R.P. bromine obtained from Prolabo, Paris without further purification. The checkers used bromine available from Fisher Scientific Company.

4. It is very important to add the bromine in a single portion. When it is added dropwise, a mixture containing significant amounts of 3-bromo-3-methyl-2-butanone is obtained.

5. The temperature must be controlled carefully, especially at the end of the addition when the reaction becomes more exothermic. If the solution becomes warm, a mixture of the two isomeric bromoketones is obtained.

6. If a slight excess of bromine has been added, a light yellow color remains after reaction of one equivalent since dibromination is very slow under these conditions.

7. The quantity of water added is such that the brominated products do not separate from the aqueous methanol.

8. The water is added in order to hydrolyze the α-bromodimethyl ketals that have been produced during the reaction. The ease of hydrolysis of these bromoketals depends on the structure of the ketone. With acetylcyclohexane (1-cyclohexylethanone) or acetylcyclopentane, stirring with water for 10 minutes is sufficient for complete hydrolysis. In contrast, with phenylacetone (1-phenylethanone) or methyl ethyl ketone (2-butanone), after dilution with water, the addition of 10 equivalents of concentrated sulfuric acid with respect to ketone and stirring for 15 hours at room temperature are necessary for complete hydrolysis.

9. The submitters state that the hydrobromic acid can also be neutralized before extraction by adding 75 g. of potassium carbonate (6 g. excess) in small portions.

10. Under these extraction conditions, the ether solution contains

significant amounts of water and methanol which cannot be removed efficiently with anhydrous sodium sulfate.

11. In the crude product the ratio of 1-bromo-3-methyl-2-butanone to 3-bromo-3-methyl-2-butanone is estimated by proton magnetic resonance to be 95:5. The spectral properties of the two isomers are as follows; 1-bromo-3-methyl-2-butanone: (chloroform-d) δ, multiplicity, number of protons, assignment, coupling constant J in Hz.: 1.17 (doublet, 6, 2 × CH_3, $J = 6.9$), 3.02 (multiplet, 1, CH), 4.10 (singlet, 2, CH_2); 3-bromo-3-methyl-2-butanone: (chloroform-d) δ, multiplicity, number of protons, assignment: 1.89 (singlet, 6, 2 × CH_3), 2.46 (singlet, 3, $COCH_3$).

3. Discussion

Ketones monobrominated at the less substituted carbon are not readily prepared by simple bromination of unsymmetrical ketones, since substitution occurs mainly at the more substituted carbon.[2] While the action of hydrobromic acid on diazoketones has long been the only method of preparing bromomethyl ketones,[3] it has recently been shown that bromination of unsymmetrical ketals, (e.g., dioxolanes or dimethyl ketals) occurs to a greater extent on the less substituted carbon atom, and this constitutes an efficient route to the corresponding α-bromo ketones.[4–6] Direct bromination of 2-substituted cyclohexanones[5] and various methyl ketones[7] in methanol leads to the same result.

This procedure, in contrast to previous methods, comprises only one step and is readily adapted to large-scale preparative work. Furthermore dibromination is very slow in methanol and hence the crude reaction products contain only traces of dibromo ketones. This contrasts with the behavior in other solvents such as ether or carbon tetrachloride, where larger amounts of dibromo ketones are always present, even when one equivalent of bromine is used. Methanol is thus recommended as a brominating solvent even when no orientation problem is involved. It should be noted that α-bromomethyl ketals are formed along with α-bromoketones and must be hydrolyzed during the workup (Note 8).[7]

The regiospecificity of bromination depends on the structure of the ketone.[7] This regiospecificity is very high for methyl ketones when the α′-position is tertiary, and not as high when it is secondary. For example, acetylcyclohexane and acetylcyclopentane lead to crude products containing 100% and 85%, respectively, of bromomethyl ketone, while 2-methylcyclohexanone, methyl ethyl ketone, and phenylacetone give 65%,[5] 70%,[7] and 40%,[7] respectively, of ketone brominated at the less substituted carbon. In these latter cases, bromination of the corresponding dimethyl ketal in methanol affords better yields of these bromo ketones.[7]

1. Organic Chemistry of Hormones Laboratory, College of France, 75231, Paris, Cedex 05.
2. J. R. Catch, D. F. Elliott, D. H. Hey, and E. R. H. Jones, *J. Chem. Soc. London*, 272 (1948); J. R. Catch, D. H. Hey, E. R. H. Jones, and W. Wilson, *J. Chem. Soc. London*, 276 (1948); H. M. E. Cardwell and A. E. H. Kilner, *J. Chem. Soc. London*, 2430 (1951).
3. D. A. Clibbens and M. Nierenstein, *J. Chem. Soc. London*, **107**, 1491 (1915); J. R. Catch, D. F. Elliott, D. H. Hey, and E. R. H. Jones, *J. Chem. Soc. London*, 278 (1948).
4. A. Marquet, M. Dvolaitzky, H. B. Kagan, L. Mamlok, C. Ouannes, and J. Jacques, *Bull. Soc. Chim. Fr.*, 1822 (1961).
5. E. W. Garbisch, Jr., *J. Org. Chem.*, **30**, 2109 (1965).
6. M. Gaudry and A. Marquet, *Bull. Soc. Chim. Fr.*, 1849 (1967); *Bull. Soc. Chim. Fr.*, 4169 (1969).
7. M. Gaudry and A. Marquet, *Tetrahedron*, **26**, 5611, 5617 (1970).

2-BROMOHEXANOYL CHLORIDE

$$\text{CH}_3\text{CH}_2\text{CH}_2\text{CH}_2\text{CH}_2\text{COOH} \xrightarrow[\text{(2) (CH}_2\text{CO)}_2\text{NBr, HBr, 85}^\circ]{\text{(1) SOCl}_2,\ 65^\circ} \text{CH}_3\text{CH}_2\text{CH}_2\text{CH}_2\underset{\underset{\text{Br}}{|}}{\text{CH}}\text{COCl}$$

Submitted by DAVID N. HARPP, L. Q. BAO,
CHRISTOPHER COYLE, JOHN G. GLEASON, and SHARON HOROVITCH[1]
Checked by JAMES E. KLECKNER and ROBERT E. IRELAND

1. Procedure

Caution! This reaction should be conducted in a good hood since hydrogen chloride and bromine vapors are evolved.

A 200-ml., round-bottomed flask equipped with a magnetic stirring bar is charged with 11.6 g. (0.1 mole) of hexanoic acid (Note 1) and 10 ml. of carbon tetrachloride. After 28.8 ml. (0.4 mole) of thionyl chloride (Note 2) is added to the solution, an efficient reflux condenser with an attached drying tube is fitted to the flask. The solution is then stirred and heated with an oil bath at 65° for 30 minutes (Note 3).

The flask is removed from the oil bath and cooled to room temperature. To the reaction mixture are added successively 21.4 g. (0.12 mole) of finely powdered N-bromosuccinimide (1-bromo-2,5-pyrrolidinedione) (Note 4), 50 ml. of carbon tetrachloride, and 7 drops of aqueous 48% hydrogen bromide (Note 5). The flask is heated at 70° for 10 minutes (Note 6), and the temperature of the bath is then increased to 85° until the color of the reaction becomes light yellow (*ca.* 1.5 hours; Note 7). The reaction mixture is cooled to room temperature, and the carbon tetrachloride and excess thionyl chloride are removed under reduced pressure (Note 8). The residue is suction filtered, and the solid (Note 9) is washed several times with carbon tetrachloride (total 20 ml.) and the combined filtrate collected in a 50-ml. flask. The solvent is removed from the solution as before, and the residue is distilled into a dry ice-cooled receiver (short-path column) to give, after a small forerun, 16.1–17.1 g. (76–80%) of 2-bromohexanoyl chloride, b.p. 44–47° (1.5 mm.) as a clear, slightly yellow oil, n^{22} D 1.4707. This material is of sufficient purity for most synthetic purposes (Note 10).

The decolorization of the yellow product (Note 11) is achieved by dissolving the product in an equal volume of carbon tetrachloride (*ca.* 12 ml.) and vigorously shaking the solution thus obtained with 1.5 ml. of a freshly prepared aqueous 35% sodium thiosulfate. The two layers are completely separated after 5 minutes. The colorless bottom layer is drawn off into a 50-ml. Erlenmeyer flask. The top layer is extracted three times with 1.5 ml. of carbon tetrachloride. The combined carbon tetrachloride solution is dried over 0.5 g. (Note 12) of anhydrous magnesium sulfate for 30 minutes. The solution is then filtered into a 50-ml. distilling flask, and the magnesium sulfate is washed several times with carbon tetrachloride (total 5 ml.). The solvent is removed, and the colorless product is distilled as described above, affording 14.7–15.8 g. (69–74% overall, based on hexanoic acid; 88–92% for the decolorization step) of colorless 2-bromohexanoyl chloride, b.p. 45–47° (1.5 mm.), n^{22} D 1.4706 (Note 13), d_4^{24} 1.4017 (Notes 14 and 15).

2. Notes

1. Practical-grade hexanoic acid is obtainable from Matheson Coleman and Bell. The submitters report slightly higher yields using purified-grade hexanoic acid obtained from Fisher Scientific Company.

2. Thionyl chloride was obtained from Anachemia Chemicals Ltd.,

Fisher Scientific Company (reagent-grade) or Matheson Coleman and Bell. The first two were slightly yellow, and the latter was colorless; however, the yields of final product were identical with each brand. The excess thionyl chloride serves as a drying agent for the hexanoic acid and as a solvent for the N-bromosuccinimide, which is not very soluble in carbon tetrachloride.

3. Proton magnetic resonance spectral analysis indicates complete conversion to the acid chloride. This may be monitored by following the disappearance of the triplet (CH_2CO_2H) at δ 2.40 and the emergence of a new triplet (CH_2COCl) at δ 2.87.

4. N-Bromosuccinimide was obtained from Matheson Coleman and Bell or Aldrich Chemical Company, Inc. Product yields were optimized using 20% excess, although only 5–10% yield reductions were noted using 5% excess reagent. Recrystallizing the reagent prior to use had no noticeable effect on the overall yield of product.

5. Aqueous 48% hydrogen bromide was obtained from Baker and Adamson. Without added hydrogen bromide, the reaction was much slower.

6. If the reaction was heated too rapidly to 85°, vigorous foaming resulted.

7. Initially the reaction mixture is dark red, and there is bromine vapor in the condenser. Toward the end of the reaction the color lightens considerably and after a short period (ca. 15 minutes) begins to darken again. The heat should be removed when this darkening commences. On standing, the yellow solution may also turn black, but the yield of the product is not noticeably affected. When the stirring is stopped, succinimide (2,5-pyrrolidinedione) floats to the top of the solution. The reaction may be conveniently monitored by following the disappearance of the triplet resonance (CH_2COCl) at δ 2.87 and appearance of a triplet ($CHBrCOCl$) at δ 4.54 in the proton magnetic resonance spectrum.

8. The evaporation of solvents under reduced pressure should be performed carefully with vigorous stirring at room temperature. An oil pump protected with a dry ice trap and equipped with a manometer is used. Initially the pressure should be adjusted to prevent excessive foaming; it is reduced progressively to approximately 5 mm.

9. About 10 g. of the solid (succinimide) is collected.

10. The infrared and proton magnetic resonance spectra are identical with those of colorless, doubly distilled material; n^{22} D 1.4706.

11. When the decolorization procedure was carried out before the first distillation, inconsistent yields were obtained. About 2.5 ml. of a dark viscous liquid (giving a violet solution on dilution in carbon tetrachloride) remained in the distillation flask.

12. When more drying agent was employed, the product yield was lower.

13. A central fraction had n^{22} D 1.4704.

14. Elemental analysis for $C_6H_{10}BrClO$ was as follows; *calculated*: C, 33.75; H, 4.72; Br, 37.42; Cl, 16.60; *found*: C, 33.42; H, 4.77; Br, 37.29; Cl, 16.74. The product has the following spectral properties; infrared (NaCl) cm.$^{-1}$: 2955, 2925, 1785, 1470; proton magnetic resonance δ, multiplicity, number of protons: 4.54 (triplet, 1), 2.10 (multiplet, 2), 1.43 (multiplet, 4), 0.94 (multiplet, 3); mass spectrum m/e: 179, 177 (M–Cl).

15. The corresponding α-bromo acid is prepared by the following procedure. A 500-ml., round-bottomed flask is charged with 10.28 g. (0.048 mole) of 2-bromohexanoyl chloride and 92 ml. of acetone. The flask is then fitted with a magnetic stirring bar, a thermometer, and a 200-ml. dropping funnel in which is placed 115 ml. of aqueous saturated sodium bicarbonate (*ca.* 0.115 mole). The flask is cooled to approximately 10°, while the base is added over a period of about 45 minutes. The mixture is then acidified with concentrated hydrochloric acid. An organic layer forms at the top and is separated from the aqueous layer. The aqueous layer is extracted with three 30-ml. portions of chloroform. The combined organic extracts are dried over anhydrous magnesium sulfate. The solvent is removed under reduced pressure to give a colorless liquid. The crude yield of 2-bromohexanoic acid is 9.36 g. This product is 96% pure by gas chromatographic analysis, using a Hewlett–Packard 5750 Research Chromatograph with a 1.8 m. 4% SE-30 column at 130°, and having a flow rate of 60 ml./ minute. This product can be distilled through a short-path column, and after an 11% forerun is taken, 7.76 g. (83%) of 2-bromohexanoic acid, b.p. 64–66° (0.075 mm.), is obtained which shows one peak by gas chromatographic analysis (as above). Infrared and proton magnetic resonance spectra are consistent with the structure.

3. Discussion

The α-bromination of acids (*via* the acid chloride) has been achieved by the Hell–Volhard–Zelinsky reaction or its variances.[2] However this technique can involve reaction times of up to 2–3 days,[3] high reaction temperatures (>100°), copious evolution of HBr, and variable yields. A recent procedure,[4] while affording good overall yields, involves several steps to achieve the transformation (alkylation, proton abstraction, bromination, deacylation and deesterification).

The submitters have found the *N*-bromosuccinimide procedure to be a very general reaction. Alkyl, alicyclic, aryl and heterocyclic acetic acids have been brominated in 50–80% yield.[5] The reaction may be applied in the presence of labile benzylic hydrogens; for example, 3-phenyl-propanoic acid gives exclusively the 2-bromo-3-phenylpropanoyl chloride.[6] The procedure has several significant advantages; it is considerably faster than the known methods (overall reaction times of 2 hours are common),[7] the use of bromine is circumvented, and work-up is considerably simplified.

1. Department of Chemistry, McGill University, P.O. Box 6070, Montreal 101, Quebec, Canada.
2. H. O. House, "Modern Synthetic Reactions," 2nd ed., Benjamin, New York, 1972, pp. 477–478, and references cited therein.
3. L. A. Carpino and L. V. McAdams III, *Org. Syn.*, **50**, 31 (1970).
4. P. L. Stotter and K. A. Hill, *Tetrahedron Lett.*, 4067, (1972).
5. J. G. Gleason and D. N. Harpp, *Tetrahedron Lett.*, 3431, (1970); J. G. Gleason, unpublished results.
6. D. N. Harpp, L. Q. Bao, C. J. Black, R. A. Smith, and J. G. Gleason, *Tetrahedron Lett.*, 3235 (1974).
7. A preliminary kinetic study has shown *N*-bromosuccinimide to brominate acid chlorides more rapidly than molecular bromine.

tert-BUTYLCYANOKETENE

[2-Propenenitrile, 2-(1,1-dimethylethyl)-3-oxo]

Submitted by WALTER WEYLER, JR., WARREN G. DUNCAN,
MARGO BETH LIEWEN, and HAROLD W. MOORE[1]
Checked by B. E. SMART and R. E. BENSON

1. Procedure

A. *2,3-Dichloro-2,5-di-tert-butylcyclohex-5-ene-1,4-dione.*   A 2-l. Erlenmeyer flask is charged with 110 g. (0.5 mole) of 2,5-di-*tert*-butylbenzoquinone (2,5-di-*tert*-butyl-2,5-cyclohexadiene-1,4-dione) (Note 1) and 500 ml. of glacial acetic acid. The reaction flask is equipped with an efficient magnetic stirring bar, and chlorine gas is introduced through a safety trap into the well-stirred mixture (Note 2). During the course of the reaction the mixture becomes homogeneous and warms to about 60°. The reaction should be completed in 35–40 minutes (Note

32

3). The mixture is then flushed with nitrogen to expel excess chlorine. During this process the product crystallizes from the solution and precipitation is completed by cooling the reaction mixture to 20°. The mixture is filtered and the product is washed with 1 l. of water, and air dried to yield 112 g. (77%) of 2,3-dichloro-2,5-di-*tert*-butyl-cyclohex-5-ene-1,4-dione, m.p. 125–129° (Note 4).

B. *2-Chloro-3,6-di-tert-butyl-1,4-benzoquinone.* A 2-l. Erlenmeyer flask is charged with a solution of 112 g. (0.385 mole) of 2,3-dichloro-2,5-di-*tert*-butylcyclohex-5-ene-1,4-dione in 800 ml. of ether. A solution of 28.4 g. (0.383 mole) of diethylamine (*N*-ethylethanamine) in 50 ml. of ether is added in one portion to the vigorously swirled flask (Note 5). The reaction is instantaneous, resulting in a voluminous precipitate. The mixture is washed with two 1-l. portions of water and then with 500 ml. of aqueous saturated sodium chloride. The yellow ether solution is dried over anhydrous magnesium sulfate, the drying agent removed by filtration, and the solvent removed on a rotary evaporator to yield 96–97 g. (98–99%) of 2-chloro-3,6-di-*tert*-butyl-1,4-benzo-quinone as a yellow oil which is used without further purification (Note 6).

C. *2,3,5-Trichloro-3,6-di-tert-butylcyclohex-5-ene-1,4-dione.* A 2-l. Erlenmeyer flask is charged with a solution of 96.6 g. (0.379 mole) of crude 2-chloro-3,6-di-*tert*-butyl-1,4-benzoquinone in 500 ml. of glacial acetic acid. The reaction flask is equipped with an efficient magnetic stirring bar, and chlorine gas is introduced (Note 7). The reaction is complete in 4–5 hours (Note 8). The acetic acid solution is then flushed with nitrogen to expel excess chlorine. Approximately 1 l. of water is added, and the resulting mixture is extracted with 300 ml. of dichloro-methane. The dichloromethane solution is washed three times with water, dried over anhydrous magnesium sulfate, the drying agent removed by filtration and the solvent is removed on a rotary evapora-tor to yield 116–117 g. (94%) of 2,3,5-trichloro-3,6-di-*tert*-butylcyclo-hex-5-ene-1,4-dione as a lemon yellow oil, which is used directly in the next step (Note 9).

D. *2,5-Dichloro-3,6-di-tert-butyl-1,4-benzoquinone.* A 2-l. Erlenmeyer flask is charged with a solution of 116.6 g. (0.357 mole) of 2,3,5-trichloro-3,6-di-*tert*-butylcyclohex-5-ene-1,4-dione dissolved in 800 ml. of ether. To the vigorously swirled solution is added, in one portion, 26.2 g. (0.36 mole) of diethylamine dissolved in approximately 50 ml. of ether. The reaction, which is instantaneous, results in a voluminous precipitate (Note 10). The reaction mixture is washed with two 1-l. portions of water and then with 500 ml. of aqueous saturated sodium chloride (Note 11). The ether solution is dried over anhydrous magnesium sulfate, the drying agent removed by filtration, and the solvent removed on a rotary evaporator. The crude product is a yellow semisolid (109 g.). This material is dissolved in 300 ml. of hot ethanol, and the solution cooled first to room temperature and finally to 0°. After crystallization has set in, the flask is left at −5° to −10° overnight. The product is filtered, washed with 85% ethanol, and then air dried to yield 62–70 g. (60–67%) of yellow crystalline 2,5-dichloro-3,6-di-*tert*-butyl-1,4-benzoquinone, m.p. 68–69° (Notes 12 and 13).

E. *2,5-Diazido-3,6-di-tert-butyl-1,4-benzoquinone.* A solution of 10 g. (0.0346 mole) of 2,5-dichloro-3,6-di-*tert*-butyl-1,4-benzoquinone in 375 ml. of methanol is cooled to 5–15°. To the solution is added, over 1–2 minutes, a solution of 5 g. (0.077 mole) of sodium azide in 15 ml. of water. The initially yellow solution becomes orange during addition of the azide. The flask is then cooled to −5° to −10° for at least 4 hours. The product precipitates from the solution and is collected by filtration to yield 8.3–8.8 g. (80–85%) of 2,5-diazido-3,6-di-*tert*-butyl-1,4-benzo-quinone, m.p. 88.9–90° (decomp.). This material is recrystallized at room temperature by dissolving it in a minimum amount of chloroform, filtering the solution, and adding 2 parts of 95% ethanol to the chloroform solution. The resulting solution is cooled to −5° to −10°, and the crystalline precipitate is isolated by filtration (Note 14). The recovery is 86% and no appreciable change in melting point is observed (Note 15).

F. *tert-Butylcyanoketene.* Typically the ketene is prepared by dissolving 1 g. of 2,5-diazido-3,6-di-*tert*-butyl-1,4-benzoquinone in 10–25 ml. of anhydrous benzene (Note 16). The solution is refluxed, and the disappearance of the starting material as well as the intermediate cyclopentenedione is followed by thin layer chromatography (Note 17). When the cyclopentenedione is no longer detectable, after approximately 90 minutes, the heating is stopped. The solution contains *tert*-butylcyanoketene in amounts equivalent to at least a 95% yield (Note 18).

2. Notes

1. Practical-grade 2,5-di-*tert*-butyl-1,4-benzoquinone of m.p. 151–154° obtained from Eastman Organic Chemicals was used. Chlorine

available from Air Products and Chemicals, Inc., was used by the checkers.

2. A satisfactory way to accomplish the introduction of chlorine with minimal loss of the gas is to seal the reaction flask with a two-holed stopper equipped with a gas-inlet tube reaching just above the surface of the reaction mixture and an exit tube connected to a U-tube filled with mineral oil which is used as a gas-flow indicator. Chlorine is then introduced from the cylinder through a safety trap at such a rate as to maintain a small positive pressure in the reaction flask.

3. The reaction can be followed by proton magnetic resonance spectroscopy. The original absorption for the vinyl proton disappears and two new absorption peaks appear, one in the vinyl region (*ca.* δ 6.5, chloroform-*d*) and the other in the methine region of the spectrum. There are two products formed, presumably the *cis*- and *trans*-isomers, in the ratio of 95:5, respectively. The checkers also obtained the same yield when the reaction quantities were doubled.

4. An analytical sample has a m.p. of 127.–129°. Additional product can be recovered from the mother liquor by addition of approximately 1 l. of water followed by filtration. The yield of this product is about 31 g. (21%). However it contains about 20% of the minor isomer (Note 3) that is not dehydrohalogenated under the reaction conditions employed in the next step. The second crop can be recrystallized from hot methanol giving predominantly the desired isomer. In some preparations the submitters did not separate the minor product and observed no significant loss in yield in the subsequent steps. The spectral properties of the product are as follows; infrared (Nujol) cm.$^{-1}$: 1700 (C=O), 1600 (C=C); proton magnetic resonance (chloroform-*d*) δ, multiplicity, number of protons, assignment: 1.28 [singlet, 9, C(CH_3)$_3$], 1.37 [singlet, 9, C(CH_3)$_3$], 4.75 (singlet, 1, CH), 6.47 (singlet, 1, =CH).

5. The reaction mixture warms slightly, resulting in the boiling of the ether. The large amount of diethylamine hydrochloride formed transforms the reaction mixture into a thick paste.

6. The spectral properties of the product are as follows; infrared (Nujol) cm.$^{-1}$: 1680 (C=O), 1660 (C=C); proton magnetic resonance (chloroform-*d*) δ, multiplicity, number of protons, assignment: 1.30 [singlet, 9, C(CH_3)$_3$], 1.46 [singlet, 9, C(CH_3)$_3$], 6.59 (singlet, 1, =CH).

7. The experimental setup in this reaction is exactly as that described in Note 2.

8. The progress of the reaction is followed by proton magnetic resonance spectroscopy. When the absorption for the vinyl proton (*ca.* δ 6.6, chloroform-*d*) is completely absent, the reaction is stopped. Several minor products that were not identified are also formed in this step.

9. The spectral properties of the product are as follows; infrared (Nujol) cm.$^{-1}$: 1710 (C=O); proton magnetic resonance (chloroform-*d*) δ, multiplicity, number of protons, assignment: 4.87 (singlet, 1, C*H*), 4.68 (singlet, 1, C*H*), 1.1–1.4 [multiplet, 18, 2 $\times$ C(C*H*$_3$)$_3$].

10. Comments given in Note 5 apply here also.

11. The solution may have a brown tint, partially masking the yellow color of the quinone. The dark color is probably due to reaction of the diethylamine with the 2,5-dichloro-3,6-di-*tert*-butyl-1,4-benzoquinone.

12. The mother liquor and wash solution are combined and concentrated to 200 ml. on a rotary evaporator. Upon cooling, a second crop (15–18 g.) of product is obtained. This second crop was a semisolid material. The spectral properties of the crystalline product are as follows; infrared (Nujol) cm.$^{-1}$: 1660 (C=O); proton magnetic resonance (chloroform-*d*) δ, multiplicity, assignment: 1.45 [singlet, C(C*H*$_3$)$_3$].

13. Theoretically there remains about 22% of product to be isolated. Some of this material can be recovered indirectly by converting it to the diazide. With 500 ml. of methanol the submitters diluted the mother liquor, which contains at most 23 g. (0.08 mole) of 2,5-dichloro-3,6-di-*tert*-butyl-1,4-benzoquinone, and then added, with swirling, a solution of 10.4 g. (0.16 mole) of sodium azide in 30 ml. of water over a 2-minute period. The yellow solution turns orange in color. It is cooled to $-5°$ to $-10°$, and the resulting orange precipitate is collected to yield 12 g. of the diazide. The minimum yield is thus 88%.

14. Water can be added to the mother liquor, and the mixture extracted with chloroform to increase the diazide recovery to nearly quantitative (95–98%). During the course of any purification method that might be employed the diazide should not be heated above 50°, since decomposition occurs quite noticeably at that temperature. It is best to store the pure product below $-5°$ in the dark, since it undergoes a facile photochemical rearrangement to the cyclopentenedione.

15. The spectral properties of the product are as follows; infrared (Nujol) cm.$^{-1}$: 2110 (N$_3$), 1640 (C=O); proton magnetic resonance (chloroform-*d*) δ, multiplicity, assignment: 1.31 [singlet, C(C*H*$_3$)$_3$].

16. The Discussion contains comments on the stability of *tert*-butyl-cyanoketene in various solvents.

17. Thin layer chromatography is carried out on silica gel using petroleum ether: chloroform (1:1) as eluent. The cyclopentenedione has an R_f value about half that of the diazide, and it can be detected with an ultraviolet lamp when silica gel containing fluorescent indicator is used. The ketene undoubtedly reacts with the hydroxyl groups of the silica gel and remains at the origin. The checkers found the reaction to be complete in 1.5–2 hours. The yield was established by the checkers to be $\geqslant 95\%$ by proton magnetic resonance spectroscopy by integration studies in the presence of an internal standard.

18. The submitters have not been successful in isolating *tert*-butylcyanoketene by any method. If the solvent is removed, the ketene polymerizes. The spectral properties of the product are as follows; infrared (benzene) cm.$^{-1}$: 2220 (C≡N), 2130 (C=C=O); proton magnetic resonance (benzene) δ, multiplicity, assignment: 0.75 [singlet, C(CH_3)$_3$].

3. Discussion

tert-Butylcyanoketene is stable to rapid self-condensation at room temperature in benzene solution. However it is quite reactive toward cycloaddition reactions with alkenes and cumulenes.[3,4] It reacts as a normal ketene with hydroxylic reagents such as alcohols to give the corresponding esters. These results are summarized in the chart on p. 38.

The stability of the ketene is less in nonaromatic hydrocarbon solvents than in aromatic solvents. For example, it has a half-life of more than 7 days in benzene at 25°. On the other hand, in cyclohexane at the same temperature its half-life is only a few hours.

All attempts to isolate *tert*-butylcyanoketene have failed. Either removal of the solvent or cooling the solution to low temperature (−70°) causes polymerization of the ketene. This is a very efficient process giving a white solid polymer which appears to have repeating keteneimine units. This assignment is consistent with the very strong absorption at 2140 cm.$^{-1}$ in the infrared spectrum.[5]

The method described here for the synthesis of *tert*-butylcyanoketene has marked advantages over other possible classical routes, such as dehydrohalogenation of the corresponding carboxylic acid

REACTIONS OF *tert*-BUTYLCYANOKETENE

chloride. The only other product formed is molecular nitrogen and no external catalyst (*e.g.*, triethylamine) is necessary. In fact when *tert*-butylcyanoketene is reacted with triethylamine, or when α-*tert*-butyl-α-cyanoacetyl chloride is subjected to the dehydrohalogenation condition, 1,3-di-*tert*-butyl-1,3-dicyanoallene is immediately formed and no ketene can be detected.

1,1-Dimethylpropylcyanoketene can be prepared in an analogous fashion starting from the commercially available 2,5-bis(1,1-dimethylpropyl)-1,4-benzenediol. This ketene seems to be very similar in stability and reactivity to its *tert*-butyl homolog.

Typical reactions of *tert*-butylcyanoketene are shown in the chart.

1. Department of Chemistry, University of California, Irvine, California.
2. Harold W. Moore and Walter Weyler, Jr., *J. Amer. Chem. Soc.*, **93**, 2812 (1971).
3. Harold W. Moore and Walter Weyler, Jr., *J. Amer. Chem. Soc.*, **92**, 4132 (1970).
4. Walter Weyler, Jr., Larry R. Byrd, Marjorie C. Caserio, and Harold W. Moore, *J. Amer. Chem. Soc.*, **94**, 1027 (1972).
5. H. K. Hall, Jr., E. P. Blanchard, Jr., S. C. Cherkofsky, J. B. Sieja, and W. A. Sheppard, *J. Amer. Chem. Soc.*, **93**, 110 (1971).

3-(4-CHLOROPHENYL)-5-(4-METHOXYPHENYL)ISOXAZOLE

Submitted by MATILDA PERKINS, CHARLES F. BEAM, JR.,[1,2]
MORGAN C. D. DYER,[3] and CHARLES R. HAUSER[4]
Checked by H. W. JACOBSON and R. E. BENSON

1. Procedure

Caution! This preparation should be carried out in an efficient hood.

A 2-l., three-necked, round-bottomed flask containing a magnetic stirring bar is fitted with a nitrogen-inlet tube and a 250-ml. pressure-equalizing dropping funnel to which is attached a calcium chloride tube. To the flask are added 16.96 g. (0.10 mole) of 4-chloroacetophenone oxime [1-(4-chlorophenyl)ethanone oxime] (Note 1) and 500 ml. of anhydrous tetrahydrofuran (Note 2). The flask is stoppered (Note 3) and cooled in an ice-water bath (Note 4). In the dropping funnel is placed 140 ml. (0.22 mole) of 1.6M butyllithium in hexane (Note 5), and this reagent is rapidly added dropwise to the stirred solution during a 12–15-minute period. The solution is stirred for 30 minutes after the addition is complete, and during this time the addition funnel is replaced by a similar clean one of 125-ml. capacity that is fitted with a drying tube containing calcium chloride (Note 6).

Cooling is continued. A solution of 8.31 g. (0.05 mole) of methyl anisate (methyl 4-methoxybenzoate) (Notes 7 and 8) in 100 ml. of anhydrous tetrahydrofuran is added to the stirred mixture during a 6–10-minute period, and the resulting mixture is then stirred for an additional 30 minutes (Note 9). At the end of this period, 300 ml. of aqueous $3N$ hydrochloric acid is added. The nitrogen-inlet tube is removed and replaced by a reflux condenser, and the dropping funnel is replaced by a ground-glass stopper. The ice bath is removed, and the mixture is heated under reflux for 1 hour. The flask is then cooled, and its contents are poured into a 2-l. Erlenmeyer flask, and solid sodium bicarbonate is added to the mixture until neutralization is complete (Note 10).

The resulting mixture consists of an organic phase and a lower aqueous phase containing a small amount of insoluble material. The mixture is transferred to a 2-l. separatory funnel and the phases are separated. The aqueous phase is extracted with 100 ml. of tetrahydrofuran, and the extract is combined with the original organic phase. The resulting solution is concentrated to dryness on a rotary evaporator. Approximately 150 ml. of xylene is added to the flask and the contents of the flask are heated to reflux to remove any water present as the azeotrope. The resulting hot solution is filtered rapidly through a large Büchner funnel to which light suction is applied. The volume is reduced to approximately 100 ml., and the solution is cooled in an ice bath. The tan crystals which separate are collected in a Büchner funnel and washed with 10 ml. of ice-cold xylene. The crude product is recrystallized from 150 ml. of xylene (Note 11) to yield, after drying, 7.4–7.6 g. (52–53%) of 3-(4-chlorophenyl)-5-(4-methoxyphenyl)isoxazole, m.p. 175–176° (Note 12).

2. Notes

1. 4-Chloroacetophenone oxime was prepared by a modification of the method described by Shriner, Fuson, and Curtin.[5] A mixture of 100 g. (0.65 mole) of reagent-grade 4-chloroacetophenone, 300 ml. of water, 200 ml. of aqueous 10% sodium hydroxide, 50 g. (0.72 mole) of hydroxylamine hydrochloride, and 500 ml. of ethanol is heated at reflux in a 2-l. round-bottomed flask for 2 hours. The crystals that separate on cooling in an ice bath are recovered by filtration and air dried. The product is added to approximately 1 l. of hexane, and the mixture is

heated to reflux to remove any remaining water as the azeotrope. The resulting solution is cooled to yield approximately 70–74 g. (64–68%) of 4-chloroacetophenone oxime as white crystals, m.p. 96–97°.

2. Tetrahydrofuran was obtained from E. I. du Pont de Nemours and Company. It was distilled from lithium aluminum hydride immediately before use. The submitters used reagent-grade tetrahydrofuran available from Matheson Coleman and Bell.

3. Ground-glass stoppers proved most convenient.

4. The initial reaction of butyllithium with the oxime is exothermic, and if the bath is not used, a slightly lower yield of colored product is obtained.

5. The concentration of the butyllithium obtained from Foote Mineral Company is generally close to the $1.6M$ as quoted. An exact measurement of the volume (hypodermic syringe recommended) is not necessary, but a slight excess above the stoichiometrically required amount (0.20 mole) is needed. The submitters used butyllithium available from Lithium Corporation of America, Inc.

6. The purpose of the exchange is to provide a clean funnel for the addition of the ester solution. If the funnel is not changed, the yield is slightly lower.

7. The ratio of the reagents is 2 oxime : 4 base : 1 ester and is consistent with similar procedures used for a modified Claisen condensation.[6] The yield is based on the ester. When a ratio of reagents of 1 oxime : 2 base : 1 ester was used, a yield of 21% based on the ester was obtained.

8. Methyl anisate was obtained from Eastman Organic Chemicals.

9. At least 30 minutes is required for an optimum yield of the isoxazole.

10. Care should be taken to add the sodium bicarbonate in small amounts initially in order to avoid excessive frothing. The mixture was tested with pH paper to establish that neutralization was complete.

11. The product can also be recrystallized from ethanol, but a substantially larger volume of solvent is required.

12. The product has the following spectral properties; infrared (KBr) cm.$^{-1}$: 3103 and 3006 (aromatic C—H), 2955, 2925, and 2830 (aliphatic C—H stretching), 1257 and 1032 (aromatic methyl ether), 841 and 812 (C—H out-of-plane bending of isoxazole C_4—H and 4-substituted phenyl); proton magnetic resonance (trifluoroacetic acid) δ, multiplicity, number of protons, assignment: 3.98 (singlet,

3, OCH_3), 7.00–7.27 (multiplet, 1, isoxazole C_4—H; and $2H$ from ArH), and 7.42–7.97 (multiplet, 6, ArH).

3. Discussion

This procedure has several advantages over previous methods. It provides a simple direct route to unsymmetrically substituted isoxazoles in which the substituents are unequivocally located. The method uses readily available starting materials and can be used for the synthesis of variety of substituted isoxazoles in which the substituents are nonreactive to butyllithium. Examples of products synthesized by this method[7] are given in Table I.

TABLE I

ISOXAZOLES DERIVED FROM OXIMES[7]

$$R_1 - \underset{N}{\overset{\frown}{\bigcirc}} \hspace{-0.5em} \underset{O}{\overset{R_2}{}}$$

R_1	R_2	Yield[a] (%)
C_6H_5—	C_6H_5—	59
C_6H_5—	4-$CH_3OC_6H_4$—	51
C_6H_5—	4-ClC_6H_4—	59
4-ClC_6H_4—	C_6H_5—	65
4-$CH_3OC_6H_4$—	4-ClC_6H_4—	66

[a] Yield obtained using ~2.25M butyllithium reagent.

3-(4-Chlorophenyl)-5-(4-methoxyphenyl)isoxazole has also been prepared from the dilithio derivative of 4-chloroacetophenone oxime by two other methods: (a) reaction with anisonitrile (4-methoxybenzonitrile) followed by acid-catalyzed cyclization[8] and (b) condensation of anisolyl chloride (4-methoxybenzoyl chloride) followed by acid-catalyzed cyclization.[9]

The concept of the use of dilithio reagents for the preparation of heterocyclic systems has been extended to the synthesis of 2-isoxazolin-5-ones[10] by carboxylation of a dilithio oxime, followed by cyclinzation, and the synthesis of pyrazoles from dilithiophenylhydrazones and trilithiothiohydrazones.[11,12]

1. Submitted from William Chandler Chemistry Laboratory, Lehigh ·University, Bethlehem, Pennsylvania 18015.

2. Inquiries should be addressed to: Department of Chemistry, Newberry College, Newberry, South Carolina 29108. Technical assistance by David C. Reames is acknowledged.

3. National Aeronautics and Space Administration Trainee, 1967–1969.

4. Deceased January 6, 1970.

5. R. L. Shriner, R. C. Fuson, and D. Y. Curtin, "The Systematic Identification of Organic Compounds," 5th ed., Wiley, New York, 1964, p. 289 (4th ed., 1956, p. 255).

6. C. R. Hauser, F. W. Swamer, and J. T. Adams, *Org. React.*, **8**, 113 (1954).

7. C. F. Beam, M. C. D. Dyer, R. A. Schwarz, and C. R. Hauser, *J. Org. Chem.*, **35**, 1806 (1970).

8. C. F. Beam, R. S. Foote, and C. R. Hauser, *J. Heterocycl. Chem.*, **9**, 183 (1972).

9. M. Perkins, C. F. Beam, and C. R. Hauser, submitted for publication.

10. J. S. Griffiths, C. F. Beam, and C. R. Hauser, *J. Chem. Soc. C*, 974 (1971).

11. R. S. Foote, C. F. Beam, and C. R. Hauser, *J. Heterocycl. Chem.*, **7**, 589 (1970).

12. C. F. Beam, R. S. Foote, and C. R. Hauser, *J. Chem. Soc. C*, 1658 (1971).

CYCLOBUTADIENE IN SYNTHESIS:
endo-TRICYCLO[4.4.0.0^{2,5}]DECA-3,8-DIEN-7,10-DIONE

Submitted by L. BRENER, J. S. McKENNIS, and R. PETTIT[1]
Checked by R. E. IRELAND and G. BROWN

1. Procedure

Caution! Because of the evolution of carbon monoxide, this procedure should be carried out in a well-ventilated hood.

A 500-ml., three-necked, round-bottomed flask fitted with a sealed mechanical stirrer and an outlet leading to a gas bubbler, is charged with a solution containing 4.0 g. (0.0208 mole) of cyclobutadieneiron tricarbonyl[2] [tricarbonyl(η^4-1,3-cyclobutadiene)iron] and 2.0 g. (0.0185 mole) of freshly sublimed *p*-benzoquinone (2,4-cyclohexadiene-1,4-dione) (Note 1) in 72 ml. of acetone and 8 ml. of water. To the vigorously stirred, ice-cold solution, approximately 40–42 g. of ceric ammonium nitrate [ammonium hexanitrocerate(IV)] (Note 2) is added portionwise over a period of 10–12 minutes (Note 3), until the carbon monoxide evolution has ceased. The reaction mixture is then poured into 600 ml.

of cold brine, and the resulting mixture is extracted with five 150-ml. portions of ether. The combined extracts are washed with four 250-ml. portions of water and then dried over anhydrous magnesium sulfate.

Removal of the solvent under reduced pressure affords 1.9–2.1 g. of the crude yellow adduct (Note 4). This crude material is dissolved in 8 ml. of hot dibutyl ether (70–80°) and rapidly percolated through approximately 2.0 g. of Florisil (Note 5). Cooling and then filtering the eluent affords 1.2–1.3 g. (40–44%) of yellow crystals, m.p. 77–80°. An additional recrystallization from dibutyl ether or ethyl acetate–hexane affords pale yellow crystals, m.p. 78.5–80° (Note 6).

2. Notes

1. In most cases involving syntheses using cyclobutadiene, it is advantageous to use an excess of the trapping agent, but here excess p-benzoquinone hampers isolation of the pure adduct.

2. Other oxidizing agents may be used to degrade cyclobutadieneiron tricarbonyl; in those cases in which the reactants or products are sensitive to the acidic ceric ammonium nitrate solutions, lead tetra-acetate in pyridine can be used.

3. Slower addition results in a diminished yield.

4. The proton magnetic resonance spectrum (benzene-d_6) indicated the presence of less than 5% p-benzoquinone. This material darkens upon standing even in a refrigerator; recrystallization should be performed as soon as possible.

5. It is imperative to use a hot narrow column to prevent crystallization and to avoid passage of dark material through the column. The column was made in a Liebig condenser of diameter 6 mm. The Florisil filled 14 cm. of the condenser. A temperature of 70–80° is sufficient to melt even pure product but still avoid passage of dark material.

6. The pure adduct had the following proton magnetic resonance spectrum (chloroform-d) δ, multiplicity, number of protons, assignment: 6.75 (singlet, 2, cyclohexene vinyl protons), 6.20 (multiplet, 2, cyclobutene vinyl protons), 3.5 (broad multiplet, 4, cyclobutane protons).

3. Discussion

This procedure is illustrative of the synthetic use of cyclobutadieneiron tricarbonyl[2] as a source of highly reactive cyclobutadiene. Cyclobutadiene has been employed, for example, in the synthesis of cubane, Dewar benzenes, and a variety of other systems.[3,4]

The synthesis of *endo*-tricyclo[4.4.0.0^{2,5}]deca-3,8-dien-7,10-dione and verification of its *endo*-configuration has been reported earlier.[3] This adduct is a useful starting material for the syntheses of tetracyclo-[5.3.0.0^{2,6}.0^{3,10}]deca-4,8-diene,[5] tricyclo[4.4.0.0^{2,5}]deca-3,7,9-triene,[6] *cis*, *syn*, *cis*-tricyclo[5.3.0.0^{2,6}]deca-4,8-dien-3,10-dione,[7] and 4-oxahexacyclo[5.4.0.0^{2,6}.0^{3,10}.0^{5,9}.0^{8,11}]undecane.[8]

1. Department of Chemistry, University of Texas, Austin, Texas 78712.
2. J. Henery and R. Pettit, *Org. Syn.*, **50**, 36 (1970); R. Pettit and J. Henery, *Org. Syn.*, **50**, 21 (1970); R. H. Grubbs, *J. Amer. Chem. Soc.*, **92**, 6693 (1970).
3. J. C. Barborak, L. Watts, and R. Pettit, *J. Amer. Chem. Soc.*, **88**, 1328 (1966).
4. R. Pettit, *Pure Appl. Chem.*, **17**, 253 (1968); J. C. Barborak and R. Pettit, *J. Amer. Chem. Soc.*, **89**, 3080 (1967); G. D. Burt and R. Pettit, *Chem. Commun.*, 517 (1965).
5. J. S. McKennis, L. Brener, J. S. Ward, and R. Pettit, *J. Amer. Chem. Soc.*, **93**, 4957 (1971).
6. E. Vedejs, *Chem. Commun.*, 536 (1971).
7. P. E. Eaton and S. A. Cerefice, *Chem. Commun.*, 1494 (1970).
8. J. S. Ward and R. Pettit, unpublished results.

6,7-DIMETHOXY-3-ISOCHROMANONE

(3*H*-2-Benzopyran-3-one, 1,4-dihydro-6,7-dimethoxy-)

Submitted by J. FINKELSTEIN and A. BROSSI[1]
Checked by YOSHINORI HAMADA and WATARU NAGATA

1. Procedure

A 500-ml., round-bottomed flask equipped with a mechanical stirrer, a dropping funnel, a thermometer, and a reflux condenser is charged with 49.0 g. (0.25 mole) of 3,4-dimethoxyphenylacetic acid (Note 1) and 125 ml. of acetic acid. The solution is stirred and heated at 80° on a steam bath, while 40 ml. of aqueous concentrated hydrochloric acid is added rapidly and followed immediately by 40 ml. of 37% formalin (Notes 2 and 3). The yellow solution is stirred and heated on a steam bath for 1 hour, during which time the reaction temperature reaches 90° (Note 4) and the solution assumes a dark brown color.

After cooling to room temperature the solution is poured, with stirring, into a mixture of 650 g. of chipped ice and 650 ml. of cold water. The mixture is transferred to a 2-l. separatory funnel, and the organic material is extracted with four 300-ml. portions of chloroform (Note 5). The combined chloroform extracts are washed with 250-ml. portions of aqueous 5% sodium bicarbonate until neutral (Note 6), then with two 250-ml. portions of water, and finally dried over anhydrous magnesium sulfate. The solvent is removed under reduced pressure on a rotary evaporator with a water bath up to a temperature of 55° (Note 7). The yellow, solid residue of crude 6,7-dimethoxy-3-isochromanone, melting at 95–100° and weighing 43.5–44.2 g. (83.7–85.1%), is suitable for general synthetic purposes. A purer product is obtained by recrystallization from 55 ml. of ethanol (Notes 8–10), giving 26–27.6 g. of white crystals which, after drying at 80°, melt at 106–108° (Note 11). Upon concentration of the mother liquor to a smaller volume, an additional 1.7–3.2 g. of the isochromanone, m.p. 103–105°, is obtained. The total yield is 29.2–29.3 g. (56.2–56.4%).

2. Notes

1. The 3,4-dimethoxyphenylacetic acid was purchased from Matheson Coleman and Bell. The checkers prepared material of m.p. 97–98° according to the procedure described in an earlier volume of this series.[2]

2. "Baker Analysed" reagent-grade formaldehyde solution obtained from J. T. Baker Chemical Company was used. The checkers used material purchased from Wako Pure Chemical Industries, Ltd., Japan.

3. About 30 seconds each is needed for the additions of concentrated hydrochloric acid and 37% formalin.

4. The temperature of the reaction mixture falls to 68° on addition of the reagents but rises again within 10 minutes.

5. Fisher Scientific Company Certified A.C.S. chloroform was used. The checkers used reagent-grade chloroform purchased from Wako Pure Chemical Industries, Ltd.

6. The wash solution should be neutral to litmus. The checkers observed that the pH values of the first and the second wash solutions were 5–5.5 and 7.5, respectively.

7. The viscous syrupy residue crystallizes on standing at room temperature or by addition of a small amount of methanol.

8. Anhydrous ethanol (Type 2B) was used.

9. Recrystallization can conveniently be performed with the initial syrupy residue.

10. The checkers found that a mixture of dichloromethane and ether was a more suitable solvent for crystallization of the product. The pure sample obtained from this solvent system melts at 108–109°.

11. The product has the following spectral properties; infrared (chloroform) cm.$^{-1}$: 3040, 1750, 1616, 1520, 1253, and 1118; proton magnetic resonance (chloroform-d) δ, multiplicity, number of protons, assignment: 3.63 (singlet, 2, CH_2CO_2), 3.88 (singlet, 6, 2 × OCH_3), 5.25 (singlet, 2, $OCOCH_2$), and 6.72, 6.77 (singlets, 1 each, aromatic H).

3. Discussion

The reaction is essentially that described by the submitters.[3] The procedure illustrates a convenient method for the synthesis of a type of lactone which could serve as an important intermediate in the synthesis of isoquinolones, tetrahydroisoquinolines, and isoquinoline alkaloids. Several analogous and closely related lactones have been reported.

The parent, unsubstituted isochromanone has been caused to react with a variety of aromatic amines to prepare N-substituted 1,4-dihydro-3(2H)-isoquinolones,[4] and with amines to give amides.[5] The 6,7-methylenedioxy-3-isochromanone was an intermediate in the synthesis of protopine and its allied alkaloids,[6] and for the synthesis of the berberine ring system.[7] The 6-methoxy analog was prepared as a potential intermediate in a camptothecin synthesis[8] and 8-methoxy-4,5,6,7-tetramethyl-3-isochromanone was an intermediate in the synthesis of sclerin.[9] The compound herein described was the basis of a facile synthesis of (±)-xylopinins,[10] and its reaction with hydrazine has been reported.[11]

The procedure may have considerable scope, as shown by the synthesis of a heterocyclic lactone which is an important intermediate in the synthesis of d,l-desoxycamptothecin, which on oxidation gave camptothecin.[12]

The nonaromatic lactones from cis-[13], and trans-2-hydroxymethyl-cyclohexaneacetic acid[14] were important intermediates in the synthesis of indole alkaloids.

1. Chemical Research Department, Hoffmann-La Roche Inc., Nutley, N. J. 07110.
2. H. R. Snyder, J. S. Buck, and W. S. Ide, *Org. Syn.*, Coll. Vol. **2**, 333 (1961).
3. J. Finkelstein and A. Brossi, *J. Heterocycl. Chem.*, **4**, 315 (1967).
4. Y. Shoo, E. C. Taylor, K. Mislow, and M. Raban, *J. Amer. Chem. Soc.*, **89**, 4910 (1967).
5. G. A. Swan, *J. Chem. Soc., London*, 1720 (1949).
6. T. S. Stevens, *J. Chem. Soc., London*, 178 (1927).
7. T. S. Stevens, *J. Chem. Soc., London*, 663 (1935).
8. T. A. Bryson, Abstracts of Papers, 136, Division of Organic Chemistry, 164th National Meeting, A.C.S., New York, N. Y., August 27–September 1, 1972.
9. T. Kubota, T. Tokoroyama, T. Nishikawa, and S. Maeda, *Tetrahedron Lett.*, 745 (1967).
10. W. Meise F. Zymalkowski, *Tetrahedron Lett.*, 1475 (1969).
11. G. Rosen and F. D. Popp, *Can. J. Chem.*, **47**, 864 (1969).
12. R. Volkmann, S. Danishefsky, J. Eggler, and D. M. Solomon, *J. Amer. Chem. Soc.*, **93**, 5576 (1971); S. Danishefsky, S. J. Etheredge, R. Volkmann, J. Eggler, and J. Quick, *J. Amer. Chem. Soc.*, **93**, 5575 (1971).
13. G. Stork and R. K. Hill, *J. Amer. Chem. Soc.*, **76**, 949 (1954).
14. E. E. van Tamelen and M. Shamma, *J. Amer. Chem. Soc.*, **76**, 950 (1954).

4,4'-DIMETHYL-1,1'-BIPHENYL

$$CH_3\text{—} \bigcirc \text{—Br} + Mg \xrightarrow{\text{tetrahydrofuran, reflux}} CH_3\text{—}\bigcirc\text{—MgBr}$$

$$2\,CH_3\text{—}\bigcirc\text{—MgBr} + 2\,TlBr \xrightarrow[\text{tetrahydrofuran, reflux}]{\text{benzene,}} CH_3\text{—}\bigcirc\text{—}\bigcirc\text{—}CH_3$$

$$+ 2\,Tl + 2\,MgBr_2$$

Submitted by L. F. Elsom, Alexander McKillop,[1]
and Edward C. Taylor[2]
Checked by Ronald F. Sieloff and Carl R. Johnson

1. Procedure

Caution! Thallium salts are very toxic. This preparation should be carried out in a well-ventilated hood. The operator should wear rubber gloves. For disposal of thallium wastes, see Note 1 on p. 74.

A. *(4-Methylphenyl)magnesium Bromide.* A 500-ml., three-necked, round-bottomed flask equipped with a reflux condenser protected by a drying tube, a mercury-sealed mechanical stirrer, and a 250-ml. pressure-equalizing dropping funnel fitted with a gas-inlet tube is thoroughly purged with dry nitrogen (Note 1). To the flask are added 6.25 g. (0.256 g.-atom) of magnesium turnings and 50 ml. of anhydrous tetrahydrofuran (Note 2), and the dropping funnel is charged with a solution of 42.7 g. (30.5 ml., 0.25 mole) of 4-bromotoluene (1-bromo-4-methylbenzene) (Note 3) in 100 ml. of anhydrous tetrahydrofuran. Approximately 10 ml. of the 4-bromotoluene solution is added to the flask, and the contents are stirred until the Grignard reaction commences (Note 4). When the initial vigorous reaction has subsided the remainder of the 4-bromotoluene solution is added at a rate such that the mixture refluxes gently. Generally the addition is complete at the end of 1 hour, and almost all of the magnesium has dissolved. The mixture is refluxed for a further hour and then cooled. The yield of (4-methylphenyl)magnesium bromide is about 95% (Note 5).

B. *4,4'-Dimethyl-1,1'-biphenyl.* A 1-l., three-necked, round-bottomed flask equipped with a reflux condenser protected by a drying tube, a mercury-sealed mechanical stirrer, and a gas-inlet tube is charged with 101 g. (0.356 mole) of thallium(I) bromide and 400 ml. of anhydrous benzene. The slurry is stirred vigorously while a stream of dry nitrogen is passed through the apparatus. The reflux condenser is temporarily removed, and the solution of (4-methylphenyl)magnesium bromide is added to the flask as rapidly as possible through a large filter funnel fitted with a *loose* plug of glass wool (Note 6). A black solid precipitates almost immediately from solution. The reflux condenser is replaced, and the contents of the flask are refluxed with stirring for 4 hours under a nitrogen atmosphere. The reaction mixture is then cooled, filtered, and the metallic thallium washed with 200 ml. of ether. The organic layer is washed once with 100 ml. of aqueous 0.1N hydrochloric acid and once with 100 ml. of water and then dried over anhydrous sodium sulfate.

The organic solvent is removed by distillation under reduced pressure to give 4,4'-dimethyl-1,1'-biphenyl contaminated with a small amount of bis(4-methylphenyl)thallium bromide. The crude product is dissolved in 30 ml. of benzene, and the solution is filtered through a short column of alumina (Note 7) using a total of 250 ml. of benzene as eluent. Distillation of the benzene under reduced pressure leaves 19–21 g. (80–83%) of 4,4'-dimethyl-1,1'-biphenyl as a colorless solid, m.p. 118–120° (Note 8).

2. Notes

1. Nitrogen is dried by passage through two Drechsel bottles containing concentrated sulfuric acid and potassium hydroxide pellets, respectively.

2. The preparation of anhydrous tetrahydrofuran is described by Seyferth.[3]

3. 4-Bromotoluene (purchased from Aldrich Chemical Company, Inc.) was distilled before use, b.p. 71–72° (15 mm.).

4. The Grignard reaction starts within a few minutes and should not require the use of a catalyst. If the reaction has not commenced within 5 minutes, the flask should be gently heated with a hot water bath until reaction starts.

5. The Grignard reagent may be standardized by the general procedure described for cyclohexylmagnesium chloride.[4]

6. A loose plug of glass wool prevents any unreacted magnesium metal being added to the reaction mixture. Care must be taken to ensure that the plug is loose enough to allow rapid addition of the Grignard reagent.

7. The dimensions of the alumina column are not critical. A column approximately 2.5 cm. by 12.5 cm. is recommended. The small amount of bis(4-methylphenyl)thallium bromide formed in the reaction remains on the top of the column.

8. The reaction may be conducted on two or three times the scale described with no decrease in yield.

3. Discussion

4,4'-Dimethyl-1,1'-biphenyl has been prepared by a wide variety of procedures, but few of these are of any practical synthetic utility. Classical radical biaryl syntheses such as the Gomberg reaction or the thermal decomposition of diaroyl peroxides give complex mixtures of products in which 4,4'-dimethyl-1,1'-biphenyl is a minor constituent. A radical process may also be involved in the formation of 4,4'-dimethyl-1,1'-biphenyl (13%) by treatment of 4-bromotoluene with hydrazine hydrate.[5] 4,4'-Dimethyl-1,1'-biphenyl has been obtained in moderate to good yield (68–89%) by treatment of either dichlorobis(4-methylphenyl)tellurium or 1,1'-tellurobis(4-methylbenzene) with degassed Raney nickel in 2-methoxyethyl ether.[6]

All of the useful procedures described for the preparation of 4,4'-dimethyl-1,1'-biphenyl involve coupling of either a 4-halotoluene by a metal or the corresponding Grignard reagents by a metal halide. 4-Halotoluenes can be coupled directly by treatment with lithium,[7] sodium,[8–10] magnesium,[11] or copper[12,13]; yields are, however, very low in the first three cases (5–15%) and only moderate (54–60%) when copper is employed as in the Ullmann synthesis. Bis(1,5-cycloocta-diene)nickel(0) has also been used to couple 1-iodo-4-methylbenzene and gives the biaryl in 63% yield.[14]

The present method of preparation of 4,4'-dimethyl-1,1'-biphenyl is that described by McKillop, Elsom, and Taylor.[15] It has the particular advantages of high yield and manipulative simplicity and is, moreover, applicable to the synthesis of a variety of symmetrically substituted biaryls. 3,3'- and 4,4'-Disubstituted and 3,3',4,4'-tetrasubstituted 1,1'-biphenyls are readily prepared, but the reaction fails when applied to the synthesis of 2,2'-disubstituted-1,1'-biphenyls. The submitters have effected the following conversions by the above procedure (starting aromatic bromide, product biphenyl, % yield); bromobenzene, biphenyl, 85; 1-bromo-4-methoxybenzene, 4,4'-dimethoxy-1,1'-biphenyl, 99; 1-bromo-3-methylbenzene, 3,3'-dimethyl-1,1'-biphenyl, 85; 4-bromo-1,2-dimethylbenzene, 3,3',4,4'-tetramethyl-1,1'-biphenyl, 76; 1-bromo-4-chlorobenzene, 4,4'-dichloro-1,1'-biphenyl, 73; 1-bromo-4-fluorobenzene, 4,4'-difluoro-1,1'-biphenyl, 73.

Related procedures, in which treatment of 4-methylphenylmagnes-ium halides with halides of copper(II),[16] silver(I),[17] cobalt(II),[18] or chromium(III)[19] also lead to the formation of 4,4'-dimethyl-1,1'-biphenyl, are either experimentally more difficult than, or do not give yields comparable to, the above method.

1. School of Chemical Sciences, University of East Anglia, Norwich, Norfolk NR4 7TJ, England.
2. Department of Chemistry, Princeton University, Princeton, New Jersey, 08540.
3. D. Seyferth, Org. Syn., Coll. Vol. 4, 259 (1963).
4. R. Lespieau and M. Bourguel, Org. Syn., Coll. Vol. 1, 187 (1932).
5. M. Busch and W. Schmidt, Ber., 62, 2612 (1929).
6. J. Bergman, Tetrahedron, 28, 3323 (1972).
7. J. F. Spencer and G. M. Price, J. Chem. Soc., London, 97, 385 (1910).
8. T. Zincke, Ber., 4, 396 (1871).
9. W. Louguinine, Ber., 4, 514, (1871).
10. M. Weiler, Ber., 29, 111 (1896).
11. H. Rupe and J. Bürgin, Ber., 44, 1218 (1911).
12. F. Ullmann, Justus Liebigs Ann. Chem., 332, 38 (1904).
13. J. Van Alphen, Rec. Trav. Chim. Pays-Bas, 50, 1111 (1931).
14. M. F. Semmelhack, P. M. Helquist, and L. D. Jones, J. Amer. Chem. Soc., 93, 5908 (1971).
15. A. McKillop. L. F. Elsom, and E. C. Taylor, Tetrahedron, 26, 4041 (1970).
16. E. Sakallerios and T. Kyrimis, Ber., 57B, 322 (1924).
17. J. H. Gardner and P. Borgstrom, J. Amer. Chem. Soc., 51, 3375 (1929).
18. M. S. Kharasch and E. K. Fields, J. Amer. Chem. Soc., 63, 2316 (1941).
19. G. N. Bennett and E. E. Turner, J. Chem. Soc., London, 105, 1057 (1914).

FRAGMENTATION OF α,β-EPOXYKETONES TO ACETYLENIC ALDEHYDES AND KETONES: PREPARATION OF 2,3-EPOXY-CYCLOHEXANONE AND ITS FRAGMENTATION TO 5-HEXYNAL

(7-Oxabicyclo[4.1.0]heptan-2-one and 5-Hexynal)

Submitted by DOROTHEE FELIX, CLAUDE WINTNER, and A. ESCHENMOSER[1]
Checked by ROBERT E. IRELAND and DAVID M. WALBA

1. Procedure

A. *2,3-Epoxycyclohexanone.* A 300-ml., three-necked, round-bottomed flask equipped with a magnetic stirring bar and a thermometer is charged with a solution of 9.60 g. (0.10 mole) of 2-cyclohexen-1-one (Note 1) in 100 ml. of methanol. After the solution is cooled to 1–3° in an ice bath, 30 ml. (34 g., 0.30 mole) of aqueous 30% hydrogen peroxide (Note 2) is added. The mixture is stirred vigorously and kept well cooled in the ice bath, and 0.15 ml. (0.00075 mole) of aqueous 20% sodium hydroxide is added in one portion. The temperature of the reaction mixture rises to approximately 30° within a few minutes and then falls again to 3–5°. Fifteen minutes after addition of the sodium hydroxide solution, the cold reaction mixture is poured into a 1-l. separatory funnel containing 150 g. of ice and 200 ml. of aqueous saturated sodium chloride, and the resulting suspension is extracted with 200 ml. of dichloromethane. After two further extractions of the aqueous layer with 150-ml. portions of dichloromethane, the combined organic extracts are dried over anhydrous magnesium sulfate, and the solvent is then removed by distillation through a 30-cm. Vigreux column (Note 3). Distillation of the residue through a 15-cm. Vigreux column under reduced pressure affords a forerun fraction of approximately 0.7 g., b.p. 60–75° (11 mm.) (Note 4), and then 8.4–8.6 g. (75–77%) of pure 2,3-epoxycyclohexanone (Notes 5 and 6), b.p. 75–77° (11 mm.), n^{20} D 1.4748, d_4^{20} 1.129.

B. *5-Hexynal*. To a solution of 5.60 g. (0.050 mole) of 2,3-epoxy-cyclohexanone in 120 ml. of benzene in a 500-ml. round-bottomed flask is added 10.82 g. (0.051 mole) of *trans*-1-amino-2,3-diphenylaziridine.[2] Initially, after brief swirling at room temperature, the reaction mixture is a colorless, homogeneous solution; however it rapidly turns yellow and cloudy due to separation of water. After 2 hours the benzene and water are removed as an azeotrope under reduced pressure on a rotary evaporator with the bath maintained at approximately 30°. The resulting crude mixture of diastereomeric hydrazones weighs 15.4 g. (Note 7) and is subjected directly to the fragmentation reaction (Note 8).

The fragmentation is conveniently performed in a 100-ml., wide-necked (Note 9), round-bottomed flask with a side arm. The side arm is equipped with a capillary fitted with a balloon filled with argon or nitrogen. The capillary should be set so that it does not dip into the reaction mixture. The reaction flask is provided with a magnetic stirring bar and fitted with a short assembly for distillation under reduced pressure. A dropping funnel extends through the stillhead to a level about that of the neck of the flask. The receiver is cooled in an ice bath. A trap is positioned between the receiver and the vacuum source and cooled in a 2-propanol–dry ice mixture. The reaction flask is immersed in an oil bath at 150–155°, and the apparatus is evacuated to a pressure of 11 mm. The crude hydrazone mixture is dissolved in 20 ml. of diethyl phthalate (Note 10) and carefully added in small portions from the funnel into the heated flask with the stirrer in operation. There is a rapid evolution of nitrogen, and 5-hexynal begins to distil. The time for addition of the entire hydrazone solution is approximately 2 hours, after which the funnel is washed with 5 ml. of diethyl phthalate. The temperature of the reaction flask is raised to 160–165°. The reaction is complete when there is no further evolution of nitrogen.

The last traces of product can be driven into the receiver by warming the stillhead with a heat gun.

The contents of the receiver and trap are combined with the aid of a few drops of dichloromethane and distilled through a 15-cm. Vigreux column under reduced pressure. After only a few drops of forerun, the main fraction is 2.87–3.17 g. (60–66%) of 5-hexynal, b.p. 61–62° (30 mm.), n^{20} D 1.4447, d_4^{20} 0.875 (Notes 11–13). Gas chromatographic analysis (Note 14) shows this material to contain 3–5% of unidentified impurities with longer retention times (Note 15).

2. Notes

1. 2-Cyclohexen-1-one is easily prepared by a two-step procedure from cyclohexane-1,3-dione.[3] It is available from Fluka AG CH-9470 Buchs.

2. A brand name of this reagent is Merck Perhydrol.

3. To avoid loss of the volatile epoxide, removal of the dichloromethane on a rotary evaporator is not recommended.

4. Gas chromatographic analysis at 120° on a 2.2-m. column packed with 10% diethylene glycol succinate showed the forerun fraction to contain approximately 50% product. The other fraction is pure 2,3-epoxycyclohexanone.

5. The infrared [(chloroform) 1710 cm.$^{-1}$ strong (C=O)] and ultraviolet [(ethanol) max. 298 nm. (ϵ 15)] spectra demonstrate the absence of the enone system. Proton magnetic resonance spectrum δ, multiplicity, number of protons, coupling constant J in Hz.: 1.5–3.0 (multiplet, 6), 3.23 (doublet, 1, $J = 4$), 3.60 (multiplet, 1).

6. A further 1.06 g. (9.4%) of product may be obtained by Kugelrohr distillation of the residue.

7. This crude product does not show any carbonyl absorption at 1710 cm.$^{-1}$ in the infrared spectrum due to unreacted 2,3-epoxycyclohexanone.

8. If the crude hydrazone mixture is not to be used immediately, it must be stored in a refrigerator.

9. (E)-Stilbene sublimes during the pyrolysis and may block a narrow aperture.

10. "Pract"-grade diethyl phthalate (obtainable from Fluka AG) should be redistilled at 126–129° (1 mm.) before use. It is stable under the reaction conditions and does not codistil with the product 5-hexynal.

11. 5-Hexynal is very susceptible to air oxidation.

12. The product has the following infrared spectrum cm.$^{-1}$: 3310 (C≡CH), 2725 (CH=O), 2115 (C≡C), 1720 (C=O).

13. Use of a Kugelrohr (70–80°, 30 mm.) is also satisfactory.

14. Gas chromatographic analysis was performed on a 2.2-m. 10% diethylene glycol succinate column at 80°.

15. The checkers found that gas chromatographic analysis of one sample using a 305 cm. by 0.3 cm. column packed with 10% SF-96 on Chromosorb P operated at 70° with a 60 ml./minute helium carrier gas flow rate gave five minor impurity peaks, two at shorter retention times, and three at longer retention times. None of these impurities was present in greater than 1.1%; total impurities were 3%.

3. Discussion

2,3-Epoxycyclohexanone has been prepared in 30% yield[4] by epoxidation of 2-cyclohexen-1-one with alkaline hydrogen peroxide, using a procedure described for isophorone oxide (4,4,6-trimethyl-7-oxabicyclo[4.1.0]heptan-2-one).[5] A better yield (66%) was obtained using tert-butyl hydroperoxide (1,1-dimethylethylhydroperoxide) and Triton B in benzene solution.[6] The procedure described here is simple and rapid.

The N-aminoaziridine version[7] of the α,β-epoxyketone→alkynone fragmentation is a possible alternative in situations where the simple tosylhydrazone version[8,9] fails. The tosylhydrazone method often gives good yields at low reaction temperatures, but it tends to be unsuccessful with the epoxides of enones that are not cyclic or are not fully substituted at the β-carbon atom. For example, it has been reported[9] that 2,3-epoxycyclohexanone does not produce 5-hexynal by the tosylhydrazone route. The N-aminoaziridine method can also be recommended for the preparation of acetylenic aldehydes as well as ketones.

Both trans-1-amino-2,3-diphenylaziridine and 1-amino-2-phenylaziridine give α,β-epoxyhydrazones that fragment in the desired manner between 100° and 200°, the choice of reagent being dictated by the ease of separation of the alkynone from the by-products, (E)-stilbene and styrene, respectively. The diphenylaziridine is especially useful when the alkynone is relatively volatile and easily separable by distillation from (E)-stilbene, as is the case in the present example. The phenylaziridine is less bulky and more stable to acid than the diphenyl derivative, and may be tried with sterically hindered epoxyketones. The fragmentation is often run in an inert, high-boiling solvent to reduce resinification, but in many cases it can be achieved by pyrolysis of the neat, crude hydrazone with concomitant distillation of the product.

The limitations of the reaction have not been systematically investigated, but the inherent lability of the aziridines can be expected to become troublesome in the case of epoxyketones which are slow to form hydrazones. The use of acid catalysis is curtailed by the instability of the aziridines, particularly the diphenylaziridine, in acidic media. Because of their solvolytic lability, the hydrazones are best formed in inert solvents. A procedure proven helpful in some cases is to mix the aziridine and the epoxyketone in anhydrous benzene, and then to remove the benzene on a rotary evaporator at room temperature. Water formed in the reaction is thus removed as the azeotrope. This process is repeated, if necessary, until no carbonyl band remains in the infrared spectrum of the residue.

Further examples and a mechanistic discussion may be found in reference 7. Borrevang[10] has reported a closely related fragmentation involving diazirine derivatives of cyclic α,β-epoxyketones.

1. Laboratorium für Organische Chemie, Eidgenössische Technische Hochschule, CH-8006 Zürich, Switzerland.
2. R. K. Müller, R. Joos, D. Felix, J. Schreiber, C. Wintner, and A. Eschenmoser, *Org. Syn.*, **55**, 211 (1975).
3. W. F. Gannon and H. O. House, *Org. Syn.*, Coll. Vol. **5**, 294, 539 (1973).
4. H. O. House and R. L. Wasson, *J. Amer. Chem. Soc.*, **79**, 1488 (1957).
5. R. L. Wasson and H. O. House, *Org. Syn.*, Coll. Vol. **4**, 552 (1963).
6. N. C. Yang and R. A. Finnegang, *J. Amer. Chem. Soc.*, **80**, 5845 (1958).
7. D. Felix, R. K. Müller, U. Horn, R. Joos, J. Schreiber, and A. Eschenmoser, *Helv. Chim. Acta*, **55**, 1276 (1972).
8. A. Eschenmoser, D. Felix, and G. Ohloff, *Helv. Chim. Acta*, **50**, 708 (1967); D. Felix, J. Schreiber, G. Ohloff, and A. Eschenmoser, *Helv. Chim. Acta*, **54**, 2896 (1971).
9. M. Tanabe, D. F. Crowe, R. L. Dehn, and G. Detre, *Tetrahedron Lett.*, 3739 (1967); M. Tanabe, D. F. Crowe, and R. L. Dehn, *Tetrahedron Lett.*, 3943 (1967).
10. P. Borrevang, J. Hjort, R. T. Rapala, and R. Edie, *Tetrahedron Lett.*, 4905 (1968).

FREE RADICAL CYCLIZATION:
ETHYL 1-CYANO-2-METHYLCYCLOHEXANECARBOXYLATE

(1-Cyano-2-methylcyclohexanecarboxylic acid, ethyl ester)

$$CH_3\text{---}C{=}C\text{---}H,\ (CH_2)_3OH \xrightarrow[\text{pyridine,}\ 0° \text{ to } 5°]{4\text{-}CH_3C_6H_4SO_2Cl} CH_3\text{---}C{=}C\text{---}H,\ (CH_2)_3OSO_2C_6H_4CH_3\text{-}4$$

$$\xrightarrow[(CH_3)_2NCHO,\ 100°]{Na\overset{+}{C}H(CN)COOC_2H_5} CH_3\text{---}C{=}C\text{---}H,\ (CH_2)_3CH(CN)COOC_2H_5 \xrightarrow[\substack{\text{cyclohexane,}\\ \text{reflux}}]{(C_6H_5COO)_2}$$

ring with CH$_3$, CN, COOC$_2$H$_5$

Submitted by MARC JULIA and MICHEL MAUMY[1]
Checked by EDWARD J. ZAIKO and HERBERT O. HOUSE

1. Procedure

Caution! Since hydrogen is evolved in this procedure, it should be performed in an efficient hood.

A. *Ethyl (E)-2-Cyano-6-octenoate.* A solution of 38.4 g. (0.384 mole) of (E)-4-hexen-1-ol (Note 1) in 160 ml. of anhydrous pyridine is placed in a 1-l. Erlenmeyer flask equipped with a magnetic stirring bar, and the solution is cooled in an ice bath. Over 1 hour, 91.5 g. (0.478 mole) of p-toluenesulfonyl chloride (4-methylbenzenesulfonyl chloride) (Note 2) is added portionwise and with stirring to the reaction mixture while maintaining the temperature at 0–5°. The resulting slurry is allowed to stand overnight (Note 3) in a refrigerator and is then poured into 500 g. of an ice-water mixture. The reaction mixture is extracted with four 150-ml. portions of ether, and the combined ethereal extracts are washed successively with four 150-ml. portions of aqueous 2N sulfuric acid, 50 ml. of aqueous saturated sodium bicarbonate, and two 50-ml. portions of water. The ethereal solution is then dried over anhydrous sodium sulfate, and the solvent removed on a rotary evaporator at 25° to yield 76–88 g. (78–91%) of crude (E)-4-hexen-1-yl p-toluenesulfonate as a pale yellow oil (Note 4).

A dry, 2-l., three-necked, round-bottomed flask is equipped with a sealed mechanical stirrer, a thermometer, and a pressure-equalizing dropping funnel fitted with a calcium chloride drying tube. Sodium hydride dispersion (Note 5), [19.2 g. of a 50% w/w suspension in mineral oil, 9.6 g. (0.40 mole) of sodium hydride] is placed in the flask, and 100 ml. of anhydrous pentane is added. After the dispersion has been stirred, the sodium hydride is allowed to settle and the supernatant liquid is removed with a pipet or a siphon. This washing operation is repeated with two additional 100-ml. portions of pentane, and 400 ml. of anhydrous N,N-dimethylformamide (Note 6) is then added to the reaction flask. To the vigorously stirred suspension of sodium hydride in N,N-dimethylformamide is added, dropwise over 30 minutes, 68 g. (0.60 mole) of ethyl cyanoacetate (Note 7). The mixture is stirred until a clear solution is obtained, and then all of the previously prepared crude (E)-4-hexen-1-yl p-toluenesulfonate, dissolved in 100 ml. of anhydrous N,N-dimethylformamide (Note 6), is added all at once. The resulting solution is slowly heated to 100°, with continuous stirring, over a period of 3 hours. During this time the reaction becomes dark red, and crystalline sodium p-toluenesulfonate separates. The mixture is allowed to stand overnight at room temperature and is then transferred to a 2-l., one-necked, round-bottomed flask, and most of the solvent is removed on a rotary evaporator. The residual semisolid is mixed with 700 ml. of water and is extracted with three 250-ml. portions of ether. The combined ethereal extracts are dried over anhydrous sodium sulfate and then concentrated on a rotary evaporator. The residual liquid is fractionally distilled under reduced pressure through a 12-cm. Vigreux column. After removal of a low-boiling forerun, b.p. 54–57° (0.2 mm.), containing mainly ethyl cyanoacetate, 34.5–38.4 g. (46–51%) of colorless ethyl (E)-2-cyano-6-octenoate is collected, b.p. 84–86° (0.2 mm.), n^{25} D 1.4458 (Notes 8 and 9).

B. *Ethyl 1-Cyano-2-methylcyclohexanecarboxylate*. A 2-l., three-necked, round-bottomed flask is equipped with a heating mantle, a Teflon®-coated magnetic stirring bar, a 1-l. pressure-equalizing dropping funnel fitted with a Teflon® stopcock (Note 10), a stopper, and a reflux condenser fitted with a nitrogen-inlet tube. The apparatus is flushed with nitrogen, and 200 ml. of freshly distilled cyclohexane and 0.30 g. (0.0012 mole) of dibenzoyl peroxide are added to the flask. While a static nitrogen atmosphere is maintained in the flask (Note 11), 5.00 g. (0.026 mole) of ethyl (E)-2-cyano-6-octenoate dissolved in

800 ml. of freshly distilled cyclohexane is added from the dropping funnel to the stirred refluxing dibenzoyl peroxide solution over a period of 40 hours (Note 10). Three additional 0.30-g. (0.0012-mole) portions of dibenzoyl peroxide (total: 1.2 g., 0.00496 mole; Note 11), are added at 12-hour intervals during the addition of the unsaturated ester. After the addition of the unsaturated ester and the dibenzoyl peroxide is complete, the reaction mixture is refluxed with stirring for an additional 20 hours (Note 12). The resulting colorless to pale yellow solution is concentrated on a rotary evaporator to approximately 200 ml. The solution is diluted with 250 ml. of ether and is then washed with 100 ml. of aqueous saturated iron(II) sulfate in order to destroy any unchanged dibenzoyl peroxide. The organic layer is then washed successively with two 200-ml. portions of aqueous saturated sodium bicarbonate and two 200-ml. portions of water. The organic layer is dried over anhydrous calcium chloride, and the solvent is removed on a rotary evaporator. The residual liquid is fractionally distilled under reduced pressure through a 5-cm. Vigreux column. After separation of a 0.16–0.25-g. (3–5%) forerun [b.p. 63–64° (0.2 mm.), n^{25} D 1.4550–1.4560] containing (Note 13) primarily the cyclized product, 3.68–3.82 g. (74–76%) of colorless ethyl 1-cyano-2-methylcyclohexanecarboxylate is collected, b.p. 64–75° (0.2 mm.), n^{25} D 1.4532–1.4539 (Notes 14 and 15).

2. Notes

1. The preparation of this unsaturated alcohol is described in this volume.[2]

2. The checkers employed a commercial sample of p-toluenesulfonyl-chloride (obtained from Eastman Organic Chemicals) without further purification.

3. The checkers found that a longer period of standing before isolation lowered the yield of the sulfonate ester.

4. Infrared (carbon tetrachloride) cm.$^{-1}$: 1375, 1195, 1185, 975 [(E)CH=CH], 940.

5. The submitters employed a dispersion of sodium hydride in mineral oil obtained from Prolabo, Paris. The checkers employed 17 g. of mineral oil dispersion containing 57% sodium hydride obtained from Alfa Inorganics, Inc.

6. The submitters dried commercial N,N-dimethylformamide over

anhydrous barium oxide for 5 days and then distilled the solvent at atmospheric pressure, b.p. 155°. The checkers allowed commercial N,N-dimethylformamide to stand over activated Linde 4A Molecular Sieves for several hours and then decanted the solvent and distilled it under reduced pressure, b.p. 43° (6 mm.).

7. The checkers employed commercial ethyl cyanoacetate (purchased from Eastman Organic Chemicals) without purification.

8. The checkers found this fractional distillation to be simplified if the initial crude product was first subjected to a rapid short-path distillation under reduced pressure to remove the bulk of the dialkylated material and other high molecular weight components.

9. The spectral properties of the product are as follows; infrared (carbon tetrachloride) cm.$^{-1}$: 2250 (C≡N), 1745 (C=O), 970 [(E) CH=CH]; proton magnetic resonance (carbon tetrachloride) δ, multiplicity, number of protons, assignment, coupling constant J in Hz.: 5.1–5.8 (multiplet, 2, CH=CH), 4.25 (quartet, 2, OCH_2, $J = 7$), 3.44 [triplet, 1, CH(CN), $J = 6.5$], 1.0–2.4 (multiplet, 9, CH_2 and allylic CH_3), 1.31 (triplet, 3, CH_3, $J = 7$). Gas chromatographic analysis of the checkers' product on a 5-m. column packed with ethylene glycol isophthalate on Chromosorb P operated at 188° shows 2 products in the ratio of approximately 95:5 having retention times of 33.8 minutes and 37.0 minutes, respectively.

10. Because of difficulties in adjusting an ordinary glass stopcock to avoid leakage and to maintain a drop rate that would add the unsaturated cyanoester solution over a 40-hour period, the checkers recommend the use of a funnel equipped with a Teflon® stopcock.

11. The submitters report this reaction to be a radical chain process that requires less than 0.2 mole of dibenzoyl peroxide per mole of starting material. The checkers can offer additional evidence of the radical chain nature of the reaction from their finding that the cyclization reaction is almost completely inhibited if the refluxing solution is not protected from atmospheric oxygen.

12. The reaction solution can be analyzed to establish complete consumption of the starting unsaturated cyanoester by injecting an aliquot of the reaction solution onto a 5-m. gas chromatographic column packed with ethylene glycol isophthalate suspended on Chromosorb P operated at 190°. Under these conditions the retention times of the starting unsaturated cyanoester and the cyclized cyanoester are 31.2 minutes and 28.0 minutes, respectively.

13. Gas chromatographic analysis (Note 12) of the forerun indicated the presence of ethyl 1-cyano-2-methylcyclohexanecarboxylate and minor amounts of five or more lower boiling impurities.

14. The product, which exhibits a single gas chromatographic peak (Note 12), is presumably a mixture of stereoisomers. The spectral properties of the product are as follows; infrared (carbon tetrachloride) cm.$^{-1}$: 2250 (C≡N), 1745 (C=O); proton magnetic resonance (carbon tetrachloride) δ, multiplicity, number of protons, assignment, coupling constant J in Hz.: 4.25 (quartet, 2, OCH_2, $J = 7$), 1.0–2.5 (multiplet, 9, CH and 4 × CH_2), 1.32 (triplet, 3, ethoxy CH_3, $J = 7$), 0.97 (doublet, 3, CH_3, $J = 6.2$). This latter methyl doublet at δ 0.97 is accompanied by a second weak doublet ($J = 6.8$ Hz.) at δ 1.02 that is presumably attributable to the second stereoisomer of ethyl 1-cyano-2-methyl-cyclohexanecarboxylate; mass spectrum m/e (relative intensity): 195 (M, 5), 141 (27), 136 (40), 126 (34), 123 (62), 122 (80), 108 (100), 98 (85), 95 (62), 94 (44), 82 (27), 81 (41), 70 (51), 67 (56), 55 (50), 53 (32), 42 (27), 41 (58), 39 (31).

15. The submitters report that this free radical cyclization was also effected by heating a solution of 5.00 g. (0.026 mole) of ethyl (*E*)-2-cyano-6-octenoate and 1.25 g. (0.0086 mole) of di-*tert*-butyl peroxide [bis(1,1-dimethylethyl)peroxide] in 500 ml. of freshly distilled cyclohexane at 140° in an autoclave for 30 hours. The solution was concentrated and the residue was distilled to yield 3.4 g. (68%) of ethyl 1-cyano-2-methylcyclohexanecarboxylate.

3. Discussion

Ethyl 1-cyano-2-methylcyclohexanecarboxylate has been prepared by catalytically hydrogenating the Diels–Alder adduct from butadiene and ethyl 2-cyano-2-butenoate[3] and by the procedure described in this preparation.[4,5] This procedure illustrates a general method for the preparation of alicyclic compounds by the cyclization of δ-ethylenic carbon radicals **1**.[6] Whereas the primary 5-hexen-1-yl radical **1**

(R=R′=R″=H) cyclizes to form methylcyclopentane (*via* **3**),[7] the 1-cyano-1-carboethoxy disubstituted radicals **1** (R′=CN, R″=CO$_2$C$_2$H$_5$) lead predominantly, and sometimes exclusively, to six-membered rings **2**. Monosubstituted radicals **1** (R″=H) often give mixtures of both isomers **2** and **3**.[5,6]

1. École Normale Superieure, Laboratoire de Chimie, 24, rue Lhomond, 75231 Paris Cedex 05, France.
2. R. Paul, O. Riobé, and M. Maumy, *Org. Syn.*, **55**, 62 (1975).
3. C. J. Morell and W. G. Stoll, *Helv. Chim. Acta*, **35**, 2556 (1952).
4. M. Julia, J. M. Surzur, and L. Katz, *Bull. Soc. Chim. Fr.*, 1109 (1964).
5. M. Julia and M. Maumy, *Bull. Soc. Chim. Fr.*, 2415, 2427 (1969).
6. For a recent review, see M. Julia, *Accounts Chem. Res.*, **4**, 386 (1971).
7. R. C. Lamb, P. W. Ayers, and M. K. Toney, *J. Amer. Chem. Soc.*, **85**, 3483 (1963).

(*E*)-4-HEXEN-1-OL

Submitted by RAYMOND PAUL, OLIVIER RIOBÉ, and MICHEL MAUMY[1]
Checked by EDWARD J. ZAIKO and HERBERT O. HOUSE

1. Procedure

Caution! All operations described in this procedure should be performed in an efficient hood, because toxic chlorine and bromomethane are used in Steps A and B respectively, and hydrogen is evolved in Step C.

A. 2,3-*Dichlorotetrahydropyran.* A 1-l., three-necked, round-bottomed flask fitted with a glass-inlet tube extending nearly to the bottom of the flask, a low-temperature thermometer, an exit tube attached to a calcium chloride drying tube, and a Teflon®-coated magnetic stirring bar is charged with a solution of 118 g. (1.4 moles) of dihydropyran (Note 1) in 400 ml. of anhydrous ether. While the solution is stirred continuously it is cooled to −30° with an acetone–dry ice bath, and anhydrous chlorine (Note 2) is passed through the solution. The chlorine is introduced at such a rate that the temperature of the reaction solution does not rise above −10° (Note 3). Completion of the addition process (*ca.* 1 hour) is indicated by a rapid development of a yellow color (excess chlorine) in the reaction solution and a distinct decrease in the temperature of the reaction mixture. When the addition is complete, several drops of dihydropyran are added to discharge the yellow color, and the colorless solution is stored at −30° (Note 4) until it is used in the next step.

B. 3-*Chloro-2-methyltetrahydropyran.* A dry, 4-l., three-necked, round-bottomed flask fitted with a powerful mechanical stirrer, a reflux condenser protected by a calcium chloride drying tube, and a gas-inlet tube extending nearly to the bottom of the flask is charged with 51 g. (2.10 g.-atoms) of magnesium turnings and 1.2 l. of anhydrous ether. Bromomethane (200 g., 2.15 moles) is allowed to distil (Note 5) into the continuously stirred reaction mixture at such a rate as to maintain gentle refluxing. This addition to form an ethereal solution of methylmagnesium bromide requires approximately 2 hours. The gas-inlet tube is then replaced with a dry 1-l. dropping funnel, the top of which is protected with a calcium chloride drying tube. The reaction mixture is cooled with stirring in an ice-salt bath. The cold ethereal solution of 2,3-dichlorotetrahydropyran is placed in the dropping funnel and added dropwise and with continuous stirring and cooling to the solution of methylmagnesium bromide at such a rate that refluxing of the reaction solution does not become too vigorous. When this addition is complete, the resulting slurry is refluxed with stirring for 3 hours and then cooled in an ice bath. To the resulting cold (0°), vigorously stirred suspension is slowly added 900 ml. of cold aqueous 15% hydrochloric acid. The organic layer is separated and the aqueous phase is extracted with two 200-ml. portions of ether.The combined ethereal solution is dried over anhydrous potassium carbonate and then concentrated by distillation of the ether at atmospheric pressure. The residual liquid is distilled under reduced pressure through a 12-cm. Vigreux column to separate 122–136 g. (65–72%) of a mixture of *cis*- and *trans*-3-chloro-2-methyltetrahydropyran as a colorless liquid boiling over the range 48–95° (17–18 mm.) (Note 6) that is sufficiently pure for use in the next step (Note 7).

C. (*E*)-*4-Hexen*-1-*ol*. A dry, 3-l., three-necked, round-bottomed flask fitted with a powerful mechanical stirrer, a 500-ml. dropping funnel, and a reflux condenser protected by a calcium chloride drying tube is charged with 53 g. (2.3 g.-atoms) of finely divided sodium (Note 8) and 1.2 l. of anhydrous ether. The 3-chloro-2-methyltetra-hydropyran (136 g., 1.01 moles) is added dropwise and with rapid stirring to the suspension of sodium in ether. When the reaction commences (Note 9), the reaction mixture turns blue. After the reaction has started, the remaining chloro-ether is added, dropwise and with stirring over approximately 90 minutes, at such a rate that brisk refluxing is maintained. The resulting mixture which remains dark blue throughout the addition of the chloro-ether, is refluxed with stirring for an additional hour (Note 10) and then cooled in an ice-salt bath. The cold reaction mixture is stirred vigorously, and 30 ml. of absolute ethanol is added dropwise and with caution to the reaction mixture. After the addition of ethanol is complete, 700 ml. of water is added, again dropwise and with stirring and cooling. After the organic layer has been separated, the aqueous phase is extracted with two 200-ml. portions of ether, and the combined ether extracts are dried over anhydrous potassium carbonate and then concentrated by distillation at atmospheric pressure. The residual liquid is distilled under reduced pressure through a 12-cm. Vigreux column to separate 89–94 g. (88–93%) of (*E*)-4-hexen-1-ol as a colorless liquid, b.p. 70–74° (19 mm.), n^{25} D 1.4389 (Note 11).

2. Notes

1. Dihydropyran (purchased from Eastman Organic Chemicals) was distilled before use; b.p. 84–86°.

2. The chlorine, obtained from a compressed gas cylinder, was passed through a wash bottle containing concentrated sulfuric acid before being passed into the reaction solution.

3. This addition of chlorine has been carried out at −5–0°, but the yield is slightly decreased and the time required for the addition is greatly extended.

4. 2,3-Dichlorotetrahydropyran may be stored for a few hours at 0°, but the yield in the subsequent step is decreased and partial decomposition may have occurred.

5. The checkers employed a sealed ampoule of bromomethane of b.p. 5° (obtained from Eastman Organic Chemicals) which was cooled to 0° and opened. After a boiling chip had been added to the ampoule, it was connected to the gas-inlet tube of the reaction apparatus with rubber tubing, and the ampoule was warmed in a water bath to distil the bromomethane into the reaction vessel.

6. The submitters report that the boiling point of this mixture is 50–70° (18 mm.) or 60–80° (45 mm.), $n^{19.5}$ D 1.4596. However the checkers found that the product obtained from the initial distillation should be collected over a wider range [45–90° (17–18 mm.)], because the boiling point of the final portion of the product is raised by the higher molecular weight residue that remains in the stillpot. The isomers have been isolated by fractional distillation through a 90-cm. Crismer column.[2] The physical constants for the lower boiling *trans*-isomer are b.p. 56° (23 mm.), n^{21} D 1.4543; for the higher-boiling *cis*-isomer the values are b.p. 72° (23 mm.), n^{21} D 1.4646.

7. The checkers found that a fraction, b.p. 45–71° (18 mm.), had the following spectral properties; infrared (carbon tetrachloride): no absorption in the 3300–1600 cm.$^{-1}$ region attributable to OH, C=O, or C=C vibrations; proton magnetic resonance (chloroform-*d*) δ, multiplicity, number of protons, assignment: 3.1–4.2 (multiplet, 4, CH—Cl, CH—O, and CH_2—O), 1.0–2.5 (multiplet, 7, CH_3 and 2 × CH_2). Thin layer chromatographic analysis of this fraction on silica gel plates using chloroform as eluent indicated the presence of a major component (the *cis*- and *trans*-isomers), $R_f = 0.60$, and a minor unidentified component, $R_f = 0.14$.

8. The finely divided sodium was prepared under boiling toluene or boiling xylene by agitation of the molten sodium with a Vibromixer. After the dispersion had cooled and the sodium had settled, the toluene or xylene was decanted and the finely divided sodium was washed with two portions of anhydrous ether.

9. The reaction generally commences with the addition of approximately 5% of the chloro-ether; if not, the mixture should be heated to boiling to initiate the reaction. If a much larger proportion of the chloro-ether has been added before the reaction commences, initiation of the

reaction may be too violent to control. The *cis*-isomer was found to be more reactive toward sodium than the *trans*-isomer.

10. The submitters reported that the blue color of the mixture fades during the reflux period to leave a cream-white colored reaction mixture. In the checkers' runs (performed under a nitrogen atmosphere) the blue color was not discharged until ethanol was added to the reaction mixture.

11. The product exhibits the following spectral properties; infrared (carbon tetrachloride) cm.$^{-1}$: 3620 (OH), 3330 broad (associated OH) 970 [(*E*) CH=CH]; proton magnetic resonance (chloroform-*d*) δ, multiplicity, number of protons, assignment, coupling constant J in Hz.: 5.1–5.9 (multiplet, 2, vinyl C*H*), 3.62 (triplet, 2, C*H*$_2$—O, $J = 6.5$), 3.05 (broad, 1, O*H*), 1.3–2.4 (multiplet, 4, 2 × C*H*$_2$), 1.64 (doublet of doublets, 3, allylic C*H*$_3$, $J = 1$ and $J = 5$); mass spectrum, m/e (relative intensity): 100 (M, 3), 82 (38), 67 (100), 55 (47), 54 (22), 41 (95), 39 (36), and 31 (28). The submitters report that their product exhibited a single gas chromatographic peak on several columns. Gas chromatographic analysis of the sample obtained by the checkers using a 6-m. column packed with 1,2,3-tris(β-cyanoethoxy)propane on Chromosorb P revealed one major peak having a retention time of 20.2 minutes and a second minor peak (6% of the total peak area) having a retention time of 22.6 minutes. The mass spectrum of this minor peak exhibited the following abundant peaks: m/e (relative intensity), 100 (M, 3) 82 (38), 67 (100), 55 (42), 54 (25), 41 (97), 40 (100), 39 (35), and 31 (30). Hence this minor component appears to be an isomer of the major product, (*E*)-4-hexen-1-ol.

3. Discussion

This procedure illustrates a general method for the stereoselective synthesis of (*E*)-disubstituted alkenyl alcohols. The reductive elimination of cyclic β-halo-ethers with metals was first introduced by Paul[3] and one example, the conversion of tetrahydrofurfuryl chloride [2-(chloromethyl)tetrahydrofuran] to 4-penten-1-ol, is described in an earlier volume of this series.[4] In 1947 Paul and Riobé[5] prepared 4-nonen-1-ol by this method, and the general method has subsequently been applied to obtain alkenyl alcohols with other substitution patterns.[2,6-8] (*E*)-4-Hexen-1-ol has been prepared by this method[9] and in lower yield by an analogous reaction with 3-bromo-2-methyltetrahydropyran.[10]

The coupling reaction of 2,3-dichlorotetrahydropyran and Grignard reagents, RMgBr, has been effected with a number of reagents including those where $R = CH_3$,[2] C_2H_5,[2] $CH_3CH_2CH_2$,[2] $CH_3(CH_2)_2CH_2$,[11] C_6H_5,[11] $CH_3(CH_2)_4CH_2$,[12] β-cyclohexenylethyl,[6] β-phenylethyl,[7,13,14] vinyl,[15] and ethynyl.[16]

1. Laboratoire de Synthèse et Électrochimie Organiques, Université Catholique de l'Ouest, Angers, and Laboratoire de Chimie Organique, E.S.P.C.I., 10, rue Vauquelin, 75005, Paris.
2. O. Riobé, *Ann. Chim. Paris*, **4**, 593 (1949).
3. R. Paul, *Bull. Soc. Chim. Fr.*, **53**, (4), 424 (1933).
4. L. A. Brooks and H. R. Snyder, *Org. Syn.*, Coll. Vol. **3**, 698 (1955).
5. R. Paul and O. Riobé, *C.R.H. Acad. Sci.*, **224**, 474 (1947).
6. M. Julia and F. LeGoffic, *Bull. Soc. Chim. Fr.*, 1129 (1964).
7. J. C. Chottard and M. Julia, *Bull. Soc. Chim. Fr.*, 3700 (1968).
8. W. E. Parham and H. E. Holmquist, *J. Amer. Chem. Soc.*, **76**, 1173 (1954).
9. L. Crombie and S. H. Harper, *J. Chem. Soc. London*, 1707 (1950).
10. R. C. Brandon, J. M. Derfer, and C. E. Boord, *J. Amer. Chem. Soc.*, **72**, 2120 (1950); C. L. Stevens, B. Cross, and T. Toda, *J. Org. Chem.*, **28**, 1283 (1963).
11. R. Paul, *C.R.H. Acad. Sci.*, **218**, 122 (1944).
12. R. Paul and S. Tchelitcheff, *Bull. Soc. Chim. Fr.*, **15**, 1199 (1948).
13. M. F. Ansell and M. E. Selleck, *J. Chem. Soc. London*, 1238 (1956).
14. O. Riobé, *Bull. Soc. Chim. Fr.*, 1138 (1963).
15. H. Normant, *Bull. Soc. Chim. Fr.*, 1769 (1959).
16. L. Gouin, *Ann. Chim. Paris*, **5**, 529 (1960).

17β-HYDROXY-5-OXO-3,5-*seco*-4-NORANDROSTANE-3-CARBOXYLIC ACID

Submitted by L. MILEWICH and L. R. AXELROD[1]
Checked by A. P. KING and R. E. BENSON

1. Procedure

A 3-l., three-necked, round-bottomed flask fitted with an efficient mechanical stirrer, a 100-ml. dropping funnel, and a 500-ml. dropping funnel is charged with a solution of 10.0 g. (0.030 mole) of testosterone acetate [17β-(acetyloxy)-androst-4-en-3-one] (Note 1) in 600 ml. of tert-butyl alcohol and a solution of 5.6 g. (0.041 mole) of anhydrous potassium carbonate in 150 ml. of water. To the flask are added 100 ml. of a solution prepared from 40.0 g. (0.187 mole) of sodium metaperiodate (Note 2) in 500 ml. of water, and 10 ml. of a solution prepared from 0.8 g. (0.005 mole) of potassium permanganate in 100 ml. of water. Stirring is then begun. The remaining portions of these solutions are transferred to the appropriate dropping funnels. The metaperiodate solution is added over a period of about 30 minutes, and the permanganate solution is added as needed to maintain a pink color (Note 3). Stirring is continued for an additional 1.5 hours (Note 4), and then a solution of 20.0 g. of sodium bisulfite in 40 ml. of water is slowly added (Note 5). After stirring for 20 minutes the suspension is filtered through a pad of 10 g. of Celite® filter aid (Note 6) on a coarse sintered-glass filter, and the residual cake is washed with two 50-ml. portions of tert-butyl alcohol. The filtrates are combined and concentrated by distillation (Note 7) to a volume of about 200 ml. After cooling, 25 ml. of aqueous 10% sulfuric acid is added. The resulting mixture is extracted with three 300-ml. portions of ether. The combined ethereal extracts are washed, first with two 50-ml. portions of water, then twice with solutions prepared from 5.0 g. of sodium bisulfite in 20 ml. of water, and finally twice with 50 ml. of water. Crushed ice is then added to the ethereal layer, and the mixture is extracted with three 70-ml. portions of aqueous 10% sodium hydroxide. The aqueous layers are combined and washed with 50 ml. of ether. The aqueous layer is transferred to a 2-l. separatory funnel, crushed ice is added, and 300 ml. of aqueous 10% sulfuric acid is added carefully. The separatory funnel is shaken gently, and the product separates as a gum or stringy mass (Note 8). The mixture is extracted with four 200-ml. portions of dichloromethane, and the extracts are combined and washed with three 15-ml. portions of water. The organic layer is dried over anhydrous sodium sulfate, the drying agent is removed by filtration, and the filtrate is evaporated under reduced pressure. The resulting material is triturated with 50 ml. of acetone, and after a brief heating period crystals begin to separate. The mixture is cooled and filtered to yield 6.0–6.5 g. (63–70%) of 17β-hydroxy-5-oxo-3,5-*seco*-4-norandrostane-3-carboxylic acid, m.p. 204–205° (Note 9). An additional amount of product can be obtained from the filtrate by concentration and cooling.

2. Notes

1. Testosterone acetate was obtained from Steraloids, Inc.

2. Sodium metaperiodate was purchased from J. T. Baker Chemical Company.

3. During these additions the temperature of the reaction mixture rises to about 35°.

4. At this point the reaction mixture has a pH of about 5.2.

5. When the bisulfite solution is added, the solution acquires a deep brown color and iodine fumes develop.

6. The product was purchased from Johns–Manville Corporation.

7. Evaporation is carried out under reduced pressure with a bath temperature of 75°. A deep iodine-colored fraction first distills, followed by a colorless distillate.

8. The checkers found it convenient to separate the aqueous phase from the sticky amorphous mass and then later dissolve the product in the dichloromethane solution that was used for the extractions.

9. The reported melting points are 206.5–207°,[2] 204–205.5°,[3] 200–202°,[4] 204–206°,[5] and 192°.[6] The product has the following spectral properties; infrared (KBr) cm.$^{-1}$: 3390, 1717, 1700, and 1056; proton magnetic resonance (chloroform-*d*) δ, assignment: 1.14 (C$_{19}$—CH_3) and 0.82 (C$_{18}$—CH_3), and additional broad absorptions; [α]$_D^{28}$ + 36°.

3. Discussion

17β-Hydroxy-5-oxo-3,5-*seco*-4-norandrostane-3-carboxylic acid has been prepared by ozonolysis of testosterone[2–4] or of testosterone acetate, followed by alkaline hydrolysis,[5] and by the oxidation of testosterone acetate with ruthenium tetroxide.[6]

The present procedure corresponds to the method[7] described earlier for the synthesis of 5-oxo-3,5-*seco*-4-norcholestane-3-carboxylic acid and is useful for preparing large quantities of the subject keto acid.

1. Southwest Foundation for Research and Education, Division of Biological Growth and Development, Department of Biochemistry, San Antonio, Texas 78284.
2. C. C. Bolt, *Rec. Trav. Chim. Pays-Bas*, **57**, 905 (1938).
3. F. L. Weisenborn, D. C. Remy, and T. L. Jacobs, *J. Amer. Chem. Soc.*, **76**, 552 (1954).
4. H. J. Ringold and G. Rosenkranz, *J. Org. Chem.*, **22**, 602 (1957).
5. P. N. Rao and L. R. Axelrod, *J. Chem. Soc. London*, 1356 (1965).
6. D. M. Piatak, H. B. Bhat, and E. Caspi, *J. Org. Chem.*, **34**, 112 (1969).
7. J. T. Edwards, D. Holder, W. H. Lunn, and I. Puskas, *Can. J. Chem.*, **39**, 599 (1961).

2-IODO-*p*-XYLENE

(Benzene, 2-iodo-1,4-dimethyl)

Submitted by EDWARD C. TAYLOR,[1] FRANK KIENZLE,[1] and ALEXANDER McKILLOP[2]
Checked by GORDON S. BATES and S. MASAMUNE

1. Procedure

Caution! Thallium salts are very toxic. This preparation should be carried out in a well-ventilated hood. The operator should wear rubber gloves. For disposal of thallium wastes, see Note 1 on p. 74.

A 500-ml., round-bottomed flask equipped with a magnetic stirring bar and a glass stopper is charged with 110 ml. of trifluoroacetic acid (Note 1) and 54.34 g. (0.1 mole) of solid thallium(III) trifluoroacetate (Note 2). A clear solution is obtained after 30 minutes of vigorous stirring. Upon addition of 10.6 g. (0.1 mole) of *p*-xylene (Note 3), the reaction mixture turns brown (Note 4). After vigorous stirring for

20 minutes, the trifluoroacetic acid is removed on a rotary evaporator while the bath temperature is maintained at 35°, and the residue is dissolved in 100 ml. of ether. The solvent is again evaporated, and the solid residue is dissolved in 100 ml. of ether. With ice cooling (Note 5), a solution of 33.2 g. (0.2 mole) of potassium iodide in 100 ml. of water is added all at once, and after the resulting dark suspension is stirred vigorously for 10 minutes, a solution of 3 g. of sodium bisulfite in 30 ml. of water is added (Note 6). Yellow thallium(I) iodide is removed by filtration after another 10 minutes of vigorous stirring and washed thoroughly with 150 ml. of ether. The aqueous layer is separated and extracted with two 60-ml. portions of ether. The combined ether extracts are washed once with aqueous 10% sodium hydroxide (Note 7) and twice with 20 ml. of water. After being dried over anhydrous magnesium sulfate for 1 hour, the ether is removed on a rotary evaporator. Distillation under reduced pressure yields 18.5–19.6 g. (80–84%) of pure 2-iodo-*p*-xylene, b.p. 110–113° (19 mm.) (Notes 8–10).

2. Notes

1. This chemical is available from Aldrich Chemical Company, Inc., Halocarbon Products Corporation, Allied Chemical Corporation, or Eastman Organic Chemicals.

2. Both the submitters and the checkers used thallium(III) trifluoroacetate prepared from thallium(III) oxide and trifluoroacetic acid.[3] Although this material may be purchased from Aldrich Chemical Company, Inc. and Eastman Organic Chemicals, the submitters recommend that the reagent be freshly prepared prior to use.

3. This reagent is obtainable from major chemical suppliers.

4. The submitters report that *p*-xylylthallium bis(trifluoroacetate) precipitates after 5 minutes. The checkers did not obtain this precipitate until the bulk of the solvent had been evaporated.

5. The reaction of aqueous potassium iodide and *p*-xylylthallium bis(trifluoroacetate) is exothermic and the ether boils off unless the reaction mixture is cooled.

6. The sodium bisulfite is added to reduce some free iodine formed in this reaction. Due to the presence of trifluoroacetic acid in the reaction mixture, sulfur dioxide evolves upon addition of the bisulfite. If not done in small portions, this operation may cause overflow of the reaction mixture.

7. The sodium hydroxide solution should be added slowly, since the reaction with the acidic ether extract is exothermic and may cause boiling of the ether. The ether extract should be washed with aqueous sodium hydroxide until the aqueous layer remains basic to litmus. This extraction is self-indicating; the ether turns from a bright yellow to a light brown and color appears in the aqueous phase.

8. There is usually a lower boiling fraction of 0.1–0.3 g. consisting mainly of unreacted *p*-xylene. Also there is 1.0–1.6 g. of a dark brown residue.

9. The purity of the product may be checked by gas chromatography. The submitters used a 10-m. column with 30% QF-1 on 45/60 Chrom W. The checkers used a 2-m. column of 10% UCW-98 on WAW DMCS operated at 150°.

10. The overall time needed for this preparation is less than 5 hours. The product decomposes slowly and should be refrigerated in the dark.

3. Discussion

This procedure for the synthesis of 2-iodo-*p*-xylene is slightly modified from that of Taylor and McKillop.[3] The reaction is generally applicable to a wide range of aromatic substrates,[3,4] and, with some modifications, to thiophenes. A critical feature of this synthesis is that the entering iodine substituent always replaces thallium at the same position on the aromatic ring. The great preference of the thallium electrophile for the *para*-position in activated aromatic substrates leads therefore to iodo-compounds of high isomeric purity. With substituents capable of chelating with the thallium(III) electrophile, thallation may occur by an intramolecular delivery route, resulting in exclusive *ortho*-substitution in optimum cases. Furthermore aromatic electrophilic thallation is reversible, and under conditions of thermodynamic rather than kinetic control, *meta*-substitution often predominates. The preparation of aromatic iodo-compounds *via* aryl-thallium bis(trifluoroacetate) intermediates thus possesses the additional advantage of potential orientation control.[4]

2-Iodo-*p*-xylene has been prepared by the action of potassium iodide on diazotized *p*-xylidine (2,5-dimethylbenzenamine) (21% yield),[5] from *p*-xylene with molecular iodine in concentrated nitric acid (50% yield)[6] or in ethanol–sulfuric acid in the presence of hydrogen peroxide (64% yield),[7] and with molecular iodine in glacial acetic acid–sulfuric acid in the presence of iodic acid as a catalyst (85% yield).[8]

1. Department of Chemistry, Princeton University, Princeton, New Jersey 08540.
2. School of Chemical Sciences, University of East Anglia, Norwich, Norfolk NR4 7TJ, England.
3. A. McKillop, J. D. Hunt, M. J. Zelesko, J. S. Fowler, E. C. Taylor, G. McGillivray, and F. Kienzle, *J. Amer. Chem. Soc.*, **93**, 4841 (1971).
4. E. C. Taylor, F. Kienzle, R. L. Robey, A. McKillop, and J. D. Hunt, *J. Amer. Chem. Soc.*, **93**, 4845 (1971).
5. G. T. Morgan and E. A. Coulson, *J. Chem. Soc. London*, 2203 (1929).
6. R. L. Datta and N. R. Chatterjee, *J. Amer. Chem. Soc.*, **39**, 435 (1917).
7. L. Jurd, *Australian J. Sci. Research*, **3A**, 587 (1950).; [*C.A.*, **45**, 6592i. (1951)].
8. H. O. Wirth, O. Königstein, and W. Kern, *Justus Liebigs Ann. Chem.*, **634**, 84 (1960).

METHYL 2-ALKYNOATES FROM 3-ALKYL-2-PYRAZOLIN-5-ONES: METHYL 2-HEXYNOATE

$$CH_3CH_2CH_2\overset{\overset{\displaystyle O}{\|}}{C}CH_2COOC_2H_5 + H_2NNH_2 \longrightarrow$$

$$+ C_2H_5OH$$

$$+ 2Tl(NO_3)_3 + CH_3OH \longrightarrow$$

$$CH_3CH_2CH_2C{\equiv}CCOOCH_3 + 2TlNO_3$$

$$+ 4 HNO_3 + N_2$$

Submitted by EDWARD C. TAYLOR,[1] ROGER L. ROBEY,[1] DAVID K. JOHNSON,[1] and ALEXANDER McKILLOP[2]
Checked by F. KIENZLE and A. BROSSI

1. Procedure

Caution! Thallium compounds are highly toxic.[3] *However, they may be safely handled if prudent laboratory procedures are practiced. Rubber gloves and laboratory coats should be worn and reactions should be carried out in an efficient hood. In addition, thallium wastes should be collected and disposed of separately* (Note 1).

A. 3-(1-*Propyl*)-2-*pyrazolin*-5-*one.* A 500-ml., round-bottomed flask equipped with a magnetic stirring bar and a reflux condenser is charged with 23.7 g. (0.15 mole) of ethyl 3-oxohexanoate (Note 2), 250 ml. of ethanol, and 9.8 g. (0.17 mole) of aqueous 85% hydrazine hydrate (Note 3). The mixture is stirred for 2 hours at 0° and 2 hours at reflux, and is then reduced in volume to 50–100 ml. on a rotary evaporator. The resulting suspension is cooled to 0–5° and suction filtered to give 14–16 g. (77–83%) of colorless crystals of 3-(1-propyl)-2-pyrazolin-5-one, m.p. 204–206°, which are dried for 1–2 hours over anhydrous calcium chloride and then used without further purification (Note 4).

B. *Methyl 2-Hexynoate.* A 1-l., round-bottomed flask equipped with a magnetic stirring bar and a reflux condenser is charged with 12.62 g. (0.10 mole) of 3-(1-propyl)-2-pyrazolin-5-one and 500 ml. of methanol (Note 5). To this solution, 93.20 g. (0.21 mole) of thallium(III) nitrate trihydrate (Note 6) is slowly added so as to avoid foaming. The reaction mixture is stirred for 20 minutes at room temperature and 20 minutes at reflux (Notes 7 and 8) and then reduced to approximately half its volume by evaporation on a rotary evaporator. It is then cooled to 0–5° and filtered through fluted filter paper to remove precipitated thallium-(I) nitrate. The filter cake is washed with 150 ml. of chloroform, and 250 ml. of water is added to the filtrate. The chloroform layer is then separated, and two additional extractions with 100 ml. of chloroform are carried out. The combined chloroform layers are washed once with 100 ml. of aqueous 5% sodium bicarbonate, twice with 100 ml. of water, and then dried over anhydrous magnesium sulfate. The chloroform is removed on a rotary evaporator, and the residue is filtered through a 2 cm. by 12 cm. column of 100–200 mesh Florisil (Note 9) using approximately 250 ml. of chloroform as eluent. The chloroform is removed on a rotary evaporator, and the resulting pale yellow liquid is vacuum distilled through a 19-cm. unpacked column (Note 10) to yield 8.63–9.24 g. (68–73%) of methyl 2-hexynoate, b.p. 47–50° (5 mm.), as a colorless to slightly yellow liquid (Note 11).

2. Notes

1. The submitters recommend collection of solid wastes in an appropriate solid waste container, and liquid wastes (filtrates containing thallium residues, etc.) in suitably labeled bottles or cans. For the disposal of thallium wastes, a commercial organization specializing in the disposal of toxic materials was employed. The submitters understand that the disposal procedure consists of burying thallium wastes in deep pits after covering with sand.

2. Ethyl 3-oxohexanoate is available under the name of ethyl butyrylacetate from Aldrich Chemical Company, Inc.

3. This product is available from Matheson Coleman and Bell.

4. The pyrazolinone should be colorless. If it is not, it may be washed with a minimum of ice-cold ethanol. This procedure is convenient and yields material of adequate purity for the subsequent reaction. Additional pyrazolinone may be obtained by evaporating the filtrate and recrystallizing the residue from ethanol.

5. Commercially available anhydrous methanol was used without further treatment.

6. Thallium(III) nitrate trihydrate is best prepared fresh by dissolving with stirring 200 g. (0.44 mole) of thallium(III) oxide (available from American Smelting and Refining, Denver, Colorado) in 400 ml. of concentrated nitric acid. The submitters have found the proportion of 1 g. of thallium(III) oxide to 2 ml. of nitric acid to be best. Any suspended matter is removed by suction filtration through a medium fritted-glass funnel. The filtrate is then cooled in an ice bath with mechanical stirring to yield thallium(III) nitrate trihydrate as a fine white powder. The precipitate is separated by suction filtration through a medium fritted-glass funnel, pressed as dry as possible, and dried for approximately 6 hours in a vacuum desiccator over phosphorus pentoxide and potassium hydroxide. Longer drying times result in thallium-(III) nitrate trihydrate of poorer quality. These crystals of thallium(III) nitrate trihydrate often occlude a considerable amount of nitric acid, with a consequent decrease in reactivity. To assure removal of this occluded nitric acid, the submitters recommend grinding the initially dried material to a fine powder with a mortar and pestle and then redrying in a vacuum desiccator, again over phosphorus pentoxide and potassium hydroxide, for an additional 6 hours. The resulting extremely reactive thallium(III) nitrate trihydrate should be stored in a desiccator, since it rapidly turns brown upon contact with moist air. Thallium residues may conveniently be removed with aqueous $1N$ hydrochloric acid.

7. The reaction mixture first turns muddy brown, due to the hydrolysis of thallium(III) nitrate to thallium(III) hydroxide and thallium(III) oxide, and then yellow with the separation of colorless thallium(I) nitrate.

8. The reduction of thallium(III) to thallium(I) may be followed with potassium iodide–starch paper. A drop of solution is placed on the

paper and allowed to dry. Thallium(III) gives a purple color when the paper is moistened with water, due to the oxidation of iodide to iodine by thallium(III). Thallium(I) gives a lemon-yellow color due to the formation of thallium(I) iodide.

9. This product is available from Floridin Company, Berkley Springs, West Virginia 25411. The checkers found that this filtration was not necessary.

10. Best results were obtained with an oil bath maintained at 80–85°. The bath temperature should never exceed 100°.

11. The spectral properties of the product are as follows; infrared (liquid film) cm.$^{-1}$: 2230 strong, 1718 strong, 1428 strong, 1261 strong, 1075 strong; proton magnetic resonance (neat) δ, multiplicity, number of protons, assignment, coupling constant J in Hz.: 3.68 (singlet, 3, OCH_3), 2.34 (triplet, 2, CH_2C≡C, J = 6.8), 1.63 (multiplet, 2, CH_2), 1.01 (triplet, 3, CH_3, J = 7.2). Gas chromatographic analysis may be conveniently carried out using 10% Carbowax 20M on 60/80 Diatoport S.

3. Discussion

Methyl 2-hexynoate has been prepared by the esterification of 2-hexynoic acid, which was prepared by the carboxylation of sodium hexynylide.[4] α,β-Alkynoic acids have generally been obtained by either carboxylation of metal alkynylides or by elimination reactions.[5] In particular, they have been prepared by the elimination of enol brosylates and

TABLE I[a]

METHYL 2-ALKYNOATES FROM 2-PYRAZOLIN-5-ONES
SUBSTITUTED AT POSITION 3

Substituent	Yield of Ester (%)
CH$_3$—	53
CH$_3$CH$_2$—	70
(CH$_3$)$_2$CHCH$_2$—	79
CH$_3$(CH$_2$)$_3$CH$_2$—	79
CH$_3$(CH$_2$)$_4$CH$_2$—	78
C$_6$H$_5$—	67
4–ClC$_6$H$_4$—	43

[a] Yields are for 0.01 mole reactions.

tosylates,[6] an intramolecular Wittig reaction involving triphenyl-phosphinecarbomethoxymethylene and carboxylic acid chlorides,[7] and the base-promoted elimination reaction of 3-substituted-4,4-dichloro-2-pyrazolin-5-ones.[8]

The present method[9] affords the methyl ester directly in high yields from 2-pyrazolin-5-ones, which are readily prepared in nearly quantitative yields from readily accessible β-keto-esters. In addition, the reaction is simple to carry out, conditions are mild, and the product is easily isolated in a high state of purity. A limitation of the reaction is that only the methyl ester can be made, as other alcohols have been found to give poor yields and undesirable mixtures of products. Table I illustrates other examples of the reaction.[10]

1. Department of Chemistry, Princeton University, Princeton, New Jersey 08540.
2. School of Chemical Sciences, University of East Anglia, Norwich, Norfolk NR4 7TJ, England.
3. E. C. Taylor and A. McKillop, *Accounts Chem. Res.*, **3**, 338, (1970).
4. A. O. Zoss and G. F. Hennion, *J. Amer. Chem. Soc.*, **63**, 1151 (1941).
5. T. F. Rutlege, "Acetylenic Compounds," Reinhold, New York, 32 (1968).
6. J. C. Craig, M. D. Bergenthal, I. Fleming, and J. Harley-Mason, *Angew. Chem. Int. Ed. Engl.*, **8**, 429 (1969).
7. G. Märkl, *Chem. Ber.*, **94**, 3005 (1961).
8. L. A. Carpino, P. H. Terry, and S. D. Thatte, *J. Org. Chem.*, **31**, 2867 (1966).
9. E. C. Taylor, R. L. Robey, and A. McKillop, *Angew. Chem. Int. Ed. Engl.*, **11**, 48 (1972).
10. R. L. Robey, Ph.D. Thesis, Princeton University, Princeton, New Jersey, 1972, p. 98.

METHYL NITROACETATE

$$2\ CH_3NO_2 + 2\ KOH \xrightarrow{160°} KO_2N{=}CHCOOK + NH_3 + 2\ H_2O$$

$$KO_2N{=}CHCOOK + H_2SO_4 + CH_3OH \xrightarrow{-15°} O_2NCH_2COOCH_3 + K_2SO_4 + H_2O$$

Submitted by S. ZEN, M. KOYAMA, and S. KOTO[1]
Checked by M. ANDO and G. BÜCHI

1. Procedure

A. *Dipotassium Salt of Nitroacetic Acid.* A 3-l., three-necked, round-bottomed flask equipped with a sealed mechanical stirrer, a condenser fitted with a calcium chloride drying tube, and a pressure-equalizing dropping funnel is charged with a fresh solution of 224 g. of potassium hydroxide in 112 g. of water. From the dropping funnel is added, over 30 minutes (Note 1), 61 g. (1.0 mole) of nitromethane. The reaction mixture is heated to reflux for 1 hour in an oil bath maintained at approximately 160° (Note 2). After cooling to room temperature, the precipitated crystalline product is filtered, washed several times with methanol, and dried in a vacuum desiccator under reduced pressure to yield 71.5–80.0 g. (79–88%) of the dipotassium salt of nitroacetic acid, m.p. 262° (decomp.).

B. *Methyl Nitroacetate.* A 2-l., three-necked, round-bottomed flask equipped with a sealed mechanical stirrer, a pressure-equalizing dropping funnel fitted with a calcium chloride drying tube, and a thermometer is charged with 70 g. (0.387 mole) of finely powdered dipotassium salt of nitroacetic acid (Note 4) and 465 ml. (11.6 moles) of methanol.

The reaction mixture is cooled to $-15° \pm 3°$ and 116 g. (1.16 moles) of concentrated sulfuric acid is added with vigorous stirring over approximately 1 hour at such a rate that the reaction temperature is maintained at $-15°$. The reaction mixture is allowed to warm to room temperature over a 4-hour period and to stir for another 4 hours at room temperature. The precipitate is removed by suction filtration and the filtrate is concentrated on a rotary evaporator under reduced pressure at 30–40°. The residual oil is dissolved in benzene and washed with water. The benzene layer is dried over anhydrous sodium sulfate and the benzene is removed by distillation. Further distillation under reduced pressure yields 30–32 g. (66–70%) of methyl nitroacetate, b.p. 80–82° (8 mm.), 111–113° (25 mm.) (Note 5).

2. Notes

1. The reaction mixture heats to 60–80° during the addition of nitromethane. The mixture may require external heating to maintain this temperature. The initial yellowish color begins to turn red–brown and gradually deepens as ammonia gas is liberated.

2. The reaction mixture should not be stirred mechanically during this period in order to avoid decomposition of the product.

3. This crude product is rather pure. It can and should be employed for the esterification step without further purification. Elemental analysis for $C_2HO_4NK_2$ was as follows; *calculated:* C, 13.26; H, 0.56; N, 7.73; K, 43.16%, *found:* C, 13.27; H, 0.57; N, 7.80; K, 42.68%.

This is a hygroscopic crystalline powder and should be used immediately after drying. There is a report[2] regarding an explosion of the dry dipotassium salt prepared by another method. There is no evidence that this procedure produces the same unstable impurities.

4. This must be ground into a fine powder with a mortar and pestle immediately prior to use.

5. The spectral properties of the product are as follows; infrared (neat) cm.$^{-1}$: 1776, 1760; proton magnetic resonance (chloroform-d) δ, multiplicity, number of protons, assignment: 3.83 (singlet, 3, OCH_3), 5.20 (singlet, 2, CH_2); n^{20} D 1.4260.

3. Discussion

Methyl nitroacetate has been prepared from nitromethane through the dipotassium salt of nitroacetic acid by the classical Steinkopf method[3] but in lower yield. The dipotassium salt was obtained in 45% yield. The method has been improved by Matthews and Kubler[4] but the salt must be recrystallized prior to esterification.

This procedure[5] is an improvement in that the reaction time is reduced and the yield is improved by increasing the concentration of alkali.

The acid-catalyzed esterification has been accomplished with either hydrochloric acid[3] or sulfuric acid;[6] an improvement on the Steinkopf method has been reported,[7] but the procedure lacks the simplicity of the present method.

Application of sulfuric acid as the catalyst is considered more practical for esterification because of its higher boiling point, its incompatibility with benzene, and the stability of nitroacetic acid in the reaction mixture that allows the omission of the final neutralization step.

The ethyl ester can also be prepared from ethyl acetoacetate (ethyl 3-oxobutanoate) by the method of Rodionov[8] as well as via Steinkopf's method.[3] Ethyl nitroacetate can be prepared in >70% yield[5] from the dipotassium salt, ethanol, and sulfuric acid, with the addition of anhydrous magnesium sulfate in order to avoid the Nef reaction.[9] The propyl and 2-propyl esters can also be obtained by this method.

1. School of Pharmaceutical Sciences, Kitasato University, Tokyo, Japan.
2. D. A. Little, *Chem. Eng. News.* **27**, 1473 (1949).
3. W. Steinkopf, *Ber.*, **42**, 2026, 3925 (1909); *Justus Liebigs Ann. Chem.*, **434**, 21 (1923).

4. V. E. Matthews and D. G. Kubler, *J. Org. Chem.*, **25**, 266 (1960).
5. S. Zen, M. Koyama, and S. Koto, *Kogyo Kagaku Zasshi*, **74**, 70 (1971).
6. H. Feuer, H. B. Hass, and K. S. Warren, *J. Amer. Chem. Soc.*, **71**, 3078 (1949).
7. S. Umezawa and S. Zen, *Bull. Chem. Soc. Jap.*, **36**, 1143 (1963).
8. V. M. Rodionov, E. V. Machinskaya, and V. M. Belikov, *J. Gen. Chem. USSR*, **18**, 917 (1948).
9. W. E. Noland, *Chem. Rev.*, **55**, 137 (1955).

S(-)-α-(1-NAPHTHYL)ETHYLAMINE

[S-1-Naphthalenemethanamine, α-methyl-]

$$\alpha\text{-}C_{10}H_7CH\text{—}CH_3 \quad \xrightarrow[\substack{\text{with} \\ \text{(-)-DAG.} \\ \text{acetone, reflux}}]{\text{Resolution}} \quad$$

racemate   S(-)-isomer

(-)-DAG: (-)-2,3:4,6-di-*O*-isopropylidene-2-keto-L-gulonic acid hydrate [L-*xylo*-2-hexulosonic acid, bis-*O*-(1-methylethylidine)-]

Submitted by E. MOHACSI and W. LEIMGRUBER[1]
Checked by P. E. GEORGHIOU, J. D. LOCK, JR.,
and S. MASAMUNE

1. Procedure

A. *S(-)-α-(1-Naphthyl)ethylamine.* A mixture of 58.44 g. (0.20 mole) of (-)-2,3:4,6-di-*O*-isopropylidene-2-keto-L-gulonic acid hydrate [(-)-DAG] (Note 1) and 1.7 l. of acetone (Note 2) is placed in a 3-l. Erlenmeyer flask. A boiling chip is added, and the mixture is heated to a gentle boil. To the resulting hot solution is added cautiously but rapidly, over a 1-minute period, 34.24 g. (0.20 mole) of racemic α-(1-naphthyl)ethylamine (Note 3) in 100 ml. of acetone. The mixture is allowed to stand at room temperature for approximately 4 hours. The (-)-amine (-)-DAG salt is filtered with suction, washed with 100 ml. of acetone, and dried in a vacuum oven at 60° to constant weight. The yield of the crude (-)-amine (-)-DAG salt is 73–76 g., m.p. 205–207° (decomp.) (Note 4), $[\alpha]_D^{25}$ −14.2° (*c* 1.01%, methanol). For crystallization, the crude salt and 4.2 l. of ethanol (Note 5) are placed in a 5-l. round-bottomed flask fitted with a reflux condenser and a mechanical stirrer. The mixture is stirred and heated at reflux for about 4 hours, during which time a clear solution is obtained. The condenser is then placed in a descending position and approximately 1.4 l. of the ethanol is distilled at atmospheric pressure (Note 6), and stirring is continued (Note 7) at room temperature for about 16 hours. The purified salt is obtained as white needles which are collected on a filter and dried to constant weight. The yield is 36–37 g., m.p. 216–218° (decomp.) (Note 4), $[\alpha]_D^{25}$ −17.5° (c 1.02%, methanol). For recrystallization, the crude salt and 4.0 l. of ethanol are used. Removal of 3.0 l. of the solvent yields 33.5–33.9 g. (75–76%) of pure (-)-amine (-)-DAG salt as white needles, m.p. 219–221° (decomp.) (Note 4), $[\alpha]_D^{25}$ −18.5° (c 0.90%, methanol).

To a slurry of 33.5–33.9 g. of the pure (-)-amine (-)-DAG salt in 130 ml. of water is added 56 ml. of aqueous $2N$ sodium hydroxide, and the resulting oily suspension is extracted with four 80-ml. portions of ether. The combined ether extracts are washed with 50 ml. of water and dried over anhydrous magnesium sulfate. After filtration and removal of the ether on a rotary evaporator, the crude base is distilled under reduced pressure through a 20-cm. Vigreux column (Note 8). This operation affords 10.9–11.7 g. (85–90% yield, based on the amount of the salt used) of the pure (-)-amine as a colorless liquid, b.p. 156–157° (11 mm.), n^{24} D 1.6211–1.6212, d_4^{24} 1.056, $[\alpha]_D^{25}$ −80.1° (neat), $[\alpha]_D^{25}$ −60.4° (c 10.0%, methanol), $[\alpha]_D^{25}$ −59.3° (c 0.65%, methanol) (Notes 9 and 10).

B. *Recovery of* (-)-DAG. The basic aqueous solution (about 186 ml.) obtained after removal of the (-)-amine by ether extraction is placed in a 600-ml. beaker and cooled to 0–5° with an ice bath. The solution is stirred with a magnetic stirrer and carefully acidified with aqueous $2N$ hydrochloric acid at 0–5° to approximately pH 2 (Note 11). The precipitated (-)-DAG is filtered without delay (Note 12), washed with 20 ml. of ice water, and air dried to constant weight. The yield of recovered (-)-DAG is 20.0–20.9 g. (91–94%, yield based on the amount of the salt used), m.p. 103° (decomp.), $[\alpha]_D^{25}$ −21.6° (c 2.28%, methanol) (Notes 13 and 14).

2. Notes

1. (-)-2,3:4,6-Di-O-isopropylidene-2-keto-L-gulonic acid hydrate [(-)-DAG] is available from the Commercial Development Dept., Hoffmann–La Roche Inc., Nutley, New Jersey 07110.

2. Mallinckrodt A.R.-grade acetone was used. The checkers found

that the use of 1.5 l. of acetone, as originally recommended by the submitters, resulted in an immediate precipitation of the (-)-DAG salt upon addition of the amine, inducing incomplete mixing of the two reagents.

3. Practical-grade racemic α-(1-naphthyl)ethylamine (purchased from Norse Laboratories, Inc., Santa Barbara, Calif. 93103) was distilled before use, b.p. 117–118° (2 mm.).

4. The melting point was measured in an evacuated, sealed capillary and was found to deviate slightly from this value occasionally.

5. The checkers used ethanol containing a maximum of 0.1% (w/w) of water and a maximum of 0.001% (w/w) of benzene.

6. The salt begins to crystallize toward the end of this operation.

7. It is advisable to maintain the stirring in order to avoid the formation of lumps, thus assuring a uniform product.

8. The distillation apparatus was first flushed with nitrogen, as the amine formed a white crystalline solid on contact with atmospheric carbon dioxide.

9. Reported physical constants[2,3] of the amine are: b.p. 153° (11 mm.), d_4^{25} 1.055, $[\alpha]_D^{25}$ −80.8° (neat).

10. A 100 MHz. proton magnetic resonance spectrum (chloroform-d) of the amine in the presence of an equal amount of the chiral shift reagent, tris[3-(trifluoromethylhydroxymethylene)-d-camphorato]europium(III)[4] (submitters), or in the presence of an equal amount of tris[3-(heptafluoropropylhydroxymethylene)-d-camphorato]europium-(III) (checkers), revealed that the product contained no detectable enantiomeric isomer.

11. The pH was measured with a Beckman Zeromatic pH meter.

12. (-)-DAG is unstable in aqueous acid.

13. Thin layer chromatography of (-)-DAG on Silica Gel G using the solvent system, benzene:methanol:acetone:acetic acid (70:20:5:5), shows one spot with $R_f \sim 0.7$.

14. The reported[5] m.p. is 98–99°.

3. Discussion

S(-)-α-(1-Naphthyl)ethylamine has been prepared by resolution of the racemic amine with camphoric acid in unspecified yield.[2,3]

The procedure herein presented allows the preparation of the same optically active amine in approximately 70% yield by the use of a new resolving agent, (-)-2,3:4,6-di-O-isopropylidene-2-keto-L-gulonic acid hydrate [(-)-DAG], whose utility for the resolution of a variety of amines has been thoroughly demonstrated.[6]

(-)-DAG is an ascorbic acid derivative with the following structure:

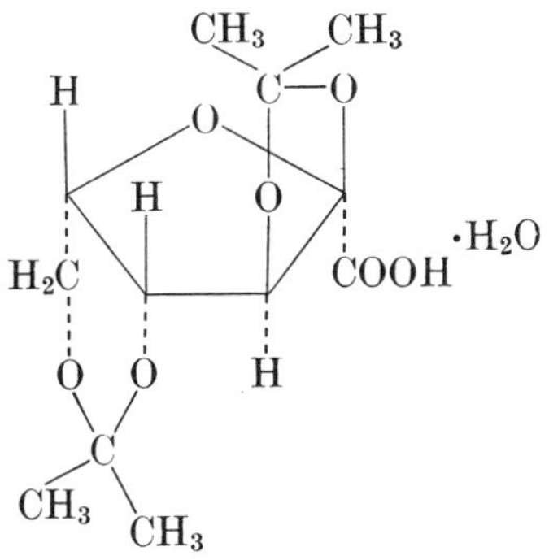

(-)-DAG

It is an attractive resolving agent, because it is relatively inexpensive and commercially available on a ton scale for industrial applications. One of the remarkable properties of (-)-DAG, which other acidic resolving agents lack, is its water-insolubility. This feature permits the recovery of the resolving agent in a simple and efficient manner.

Among the amines that have been resolved with (-)-DAG are: α-phenylethylamine (α-methylbenzenemethanamine),[6] [R-(R*, R*)]-2-amino-1-(4-nitrophenyl)-1,3-propanediol,[6] 1,2,3,4,5,6,7,8-octahydro-1-(4-methoxyphenylmethyl)isoquinoline,[6] 3-methoxymorphinan,[6] 1,2,3,4-tetrahydro-7-methoxy-4-phenylisoquinoline,[6] 3-hydroxy-N-methyl-morphinan,[6] 1,2,3,4-tetrahydro-6,7-dimethoxy-1-[3,4-(methylenedioxy)phenyl]isoquinoline,[7] 1,2,3,4-tetrahydro-6,7-dimethoxy-1-(3,4,5-trimethoxyphenyl)isoquinoline,[7] 2-[4-(phenylmethoxy)phenyl]-2-(3,4-dimethoxyphenyl)ethanamine,[8] N-norlaudanosine (tetrahydropapaverine),[9] and 1,2,3,4-tetrahydro-6,7-dimethoxy-1-[3-methoxy-4-(phenylmethoxy)phenyl]isoquinoline.[10]

1. Chemical Research Department, Hoffmann–La Roche Inc., Nutley, New Jersey 07110.
2. E. Samuelsson, *Svensk Kem. Tids.* **34**, 7 (1922) [*C.A.*, **16**, 2140 (1922)].
3. E. Samuelsson, Thesis, Univ. Lund, 1923 [*C.A.*, **18**, 1833 (1924)].
4. H. L. Goering, J. N. Eikenberry, and G. S. Koermer, *J. Amer. Chem. Soc.*, **93**, 5913 (1971).
5. T. Reichstein and A. Grüssner, *Helv. Chim. Acta*, **17**, 311 (1934).

6. C. W. Den Hollander, W. Leimgruber, and E. Mohacsi, U.S. Patent **3,682,925** (1972).

7. A. Brossi and S. Teitel, *Helv. Chim. Acta*, **54**, 1564 (1971).

8. A. Brossi and S. Teitel, *J. Org. Chem.*, **35**, 3559 (1970).

9. S. Teitel, J. O'Brien, and A. Brossi, *J. Med. Chem.*, **15**, 845 (1972).

10. J. G. Blount, V. Toome, S. Teitel, and A. Brossi, *Tetrahedron*, **29**, 31 (1973).

OXIDATION WITH THE CHROMIUM TRIOXIDE–PYRIDINE COMPLEX PREPARED *in situ*: 1-DECANAL

$$CH_3(CH_2)_8CH_2OH \xrightarrow[\text{dichloromethane, } 20°]{CrO_3 \cdot (\text{pyridine})_2} CH_3(CH_2)_8CHO$$

Submitted by R. W. RATCLIFFE[1]
Checked by ROBERT J. NEWLAND and CARL R. JOHNSON

1. Procedure

A 3-l., three-necked, round-bottomed flask equipped with a stirrer, a thermometer, and a drying tube is charged with 94.9 g. (1.2 moles) of pyridine (Note 1) and 1.5 l. of dichloromethane (Note 2). The solution is stirred with ice-bath cooling to an internal temperature of 5°, and 60.0 g. (0.6 mole) of chromium trioxide (Note 3) is added in one portion. The deep burgundy solution is stirred in the cold for an additional 5 minutes and then allowed to warm to 20° over a period of 60 minutes. A solution of 15.8 g. (0.1 mole) of 1-decanol (Note 4) in 100 ml. of dichloromethane is added rapidly with immediate separation of a tarry, black deposit. The reaction mixture is stirred for 15 minutes and then decanted from the tarry residue which is washed with three 500-ml. portions of ether. The combined organic solution is washed with three 1-l. portions of ice-cold, aqueous 5% sodium hydroxide, 1 l. of ice-cold, aqueous 5% hydrochloric acid, 1 l. of aqueous 5% sodium bicarbonate, and 1 l. of saturated brine. The solution is dried over anhydrous magnesium sulfate, filtered, and the solvents are evaporated under reduced pressure. The resulting pale yellow liquid is distilled through a 15-cm., vacuum-jacketed Vigreux column (Note 5) to give 9.8–10.2 g. (63–66%) (Note 6) of 1-decanal, b.p. 96–98° (13 mm.) (Note 7).

2. Notes

1. Anhydrous pyridine is obtained by distillation of reagent-grade material from barium oxide and storing over 4A molecular sieves.

2. Dichloromethane is purified by shaking with concentrated sulfuric acid, washing with aqueous sodium bicarbonate and water, drying over anhydrous calcium chloride, and distilling. The purified solvent is stored in the dark over 4A molecular sieves.

3. Chromium trioxide (obtained from J. T. Baker Chemical Company) is stored in a vacuum desiccator over phosphorus pentoxide prior to use. Six-mole equivalents of oxidant is required for rapid, complete conversion to aldehyde. With less than the 6:1 molar ratio, a second, extremely slow oxidation step occurs (see reference 7).

4. 1-Decanol was obtained from Aldrich Chemical Company, Inc.

5. Vigorous magnetic stirring of the pot material prevents excessive foaming during the distillation.

6. The submitters obtained 12.9–13.0 g. (83%). The checkers obtained a yield of 66% when all solvent and wash volumes used in the procedure were reduced by 50%.

7. The product was identified through comparison of its infrared, proton magnetic resonance, and mass spectra and gas chromatographic mobility with authentic 1-decanal, available from Aldrich Chemical Company, Inc.

3. Discussion

Dipyridine-chromium(VI) oxide[2] was introduced as an oxidant for the conversion of acid-sensitive alcohols to carbonyl compounds by Poos, Arth, Beyler, and Sarett.[3] The complex, dispersed in pyridine, smoothly converts secondary alcohols to ketones, but oxidations of primary alcohols to aldehydes are capricious.[4] In 1968, Collins, Hess, and Frank found that anhydrous dipyridine–chromium(VI) oxide is moderately soluble in chlorinated hydrocarbons and chose dichloromethane as the solvent.[5] By this modification, primary and secondary alcohols were oxidized to aldehydes and ketones in yields of 87–98%. Subsequently, Dauben, Lorber, and Fullerton showed that dichloromethane solutions of the complex are also useful for accomplishing allylic oxidations.[6]

The chief drawbacks to using the Collins reagent are the nuisance involved in preparing pure dipyridine–chromium(VI) oxide,[6] its hygroscopic nature[5] and its propensity to enflame during preparation.[2,3,6] The present method avoids these difficulties by simply preparing dichloromethane solutions of the complex directly.[7] In addition, as noted previously,[5] the use of dichloromethane as solvent facilitates isolation of the products.

1. Merck Sharp and Dohme Research Laboratories, Division of Merck and Company, Inc., Rahway, New Jersey 07065.
2. H. H. Sisler, J. D. Bush, and O. E. Accountius, *J. Amer. Chem. Soc.*, **70**, 3827 (1948).
3. G. I. Poos, G. E. Arth, R. E. Beyler, and L. H. Sarett, *J. Amer. Chem. Soc.*, **75**, 422 (1953).
4. J. R. Holum, *J. Org. Chem.*, **26**, 4814 (1961).
5. J. C. Collins, W. W. Hess, and F. J. Frank, *Tetrahedron Lett.*, 3363 (1968); J. C. Collins and W. W. Hess, *Org. Syn.*, **52**, 5 (1972).
6. W. G. Dauben, M. Lorber, and D. S. Fullerton, *J. Org. Chem.*, **34**, 3587 (1969).
7. R. Ratcliffe and R. Rodehorst, *J. Org. Chem.*, **35**, 4000 (1970).

1,6-OXIDO[10]ANNULENE

(11-Oxabicyclo[4.4.1]undeca-1,3,5,7,9-pentaene)

Submitted by E. Vogel, W. Klug, and A. Breuer[1]
Checked by D. Knopp, U. Schwieter, and A. Brossi

1. Procedure

A. 11-*Oxatricyclo*[4.4.1.0^{1,6}]*undeca*-3,8-*diene*. A 1-l., three-necked, round-bottomed flask equipped with a sealed mechanical stirrer, a pressure-equalizing dropping funnel, and a thermometer is charged with 66.1 g. (0.5 mole) of 1,4,5,8-tetrahydronaphthalene[2] and 200 ml. of anhydrous dichloromethane. To the resulting solution is added 75 g. of anhydrous sodium acetate. The suspension is then cooled by means of an ice bath, and 104.5 g. (0.55 mole) of commercial 40% peracetic acid (ethaneperoxoic acid) (Notes 1 and 2) is added dropwise over a period of 20–30 minutes, while maintaining the temperature at approximately 15° (Note 3) and stirring vigorously. To the reaction mixture is then added, without delay, 500 ml. of water in order to dissolve the sodium acetate and to extract the bulk of the acetic acid. The organic layer is washed successively with two 100-ml. portions of aqueous 5% sodium hydroxide and two 100-ml. portions of water and is dried over anhydrous potassium carbonate. The solvent is removed on a rotary evaporator, leaving a solid residue, which is recrystallized twice from approximately 50-ml. portions of petroleum ether (b.p. 40–60°) by cooling the solution to −40° to yield 58.5–62.0 g. (79–84%) of 11-oxatricyclo[4.4.1.0^{1,6}]undeca-3,8-diene as white needles m.p. 58–61° (Note 4).

B. 3,4,8,9,-*Tetrabromo*-11-*oxatricyclo*[4.4.1.0^{1,6}]*undecane*. A 1-l., three-necked, round-bottomed flask fitted with a sealed mechanical stirrer, a pressure-equalizing dropping funnel, and a calcium chloride drying tube is charged with 59.2 g. (0.4 mole) of 11-oxatricyclo-[4.4.1.0^{1,6}]undeca-3,8-diene and 500 ml. of anhydrous ether (Note 5). The resulting solution is cooled in an acetone–dry ice bath, and 120 g. (0.75 mole) of bromine (Note 6) is added with stirring over a period of 1.5 hours (Note 7). After the addition is complete, the ether is removed on a rotary evaporator under reduced pressure. The solid residue is dissolved in 800 ml. of hot chloroform. To this solution is added, with gentle stirring, 150 ml. of hot petroleum ether (b.p. 60–90°). The resulting clear mixture, from which the product begins to crystallize, is allowed to cool to room temperature and then to stand in a refrigerator at −40° overnight to complete the crystallization. The yield of white crystalline tetrabromide thus obtained is 115–125 g. (61–67%), m.p. 151–153° (Note 8). An additional 18–25 g. of product, m.p. 149–152°, is recovered from the mother liquor by concentration to about one quarter of the volume. The total yield is 136–144 g. (73–77%). This material is sufficiently pure to be used in the following step.

C. 1,6-*Oxido*[10]*annulene*. A 2-l., three-necked, round-bottomed flask equipped with a sealed mechanical stirrer and a reflux condenser protected by a calcium chloride drying tube is charged with 81 g. (1.5 moles) of sodium methoxide (Note 9) and 600 ml. of anhydrous ether. To this slurry is added, with stirring, 117 g. (0.25 mole) of finely powdered 3,4,8,9-tetrabromo-11-oxatricyclo[4.4.1.0^{1,6}]undecane. The reaction mixture is refluxed with stirring for 10 hours and is then allowed to stand overnight. Following this, 500 ml. of water is added slowly to dissolve the solids. The ether layer is separated and the aqueous layer is extracted with two 100-ml. portions of ether. The combined ethereal solution is washed with 250 ml. of water and dried over anhydrous potassium carbonate. Removal of the ether under reduced pressure on a rotary evaporator affords a brown oil that is distilled to give 34.4–35.1 g. of yellow 1,6-oxido[10]annulene, b.p. 77° (0.02 mm.) (Note 10). This material readily solidifies at room temperature. After two recrystallizations at −40° from 225-ml. portions of pentane–ether (5:1), 18.3–18.5 g. (51%) of 1,6-oxido[10]annulene is obtained as pale yellow needles, m.p. 51–52° (Notes 11 and 12).

2. Notes

1. Satisfactory 40% peracetic acid is obtainable from Buffalo Electrochemical Corporation, Food Machinery and Chemical Corporation, Buffalo, New York. The specifications given by the manufacturer for its composition are: peracetic acid, 40%; hydrogen peroxide, 5%; acetic acid, 39%; sulfuric acid, 1%; water, 15%. Its density is 1.15 g./ml. The peracetic acid concentration should be determined by titration. A method for the analysis of peracid solutions is based on the use of ceric sulfate as a titrant for the hydrogen peroxide present, followed by an iodometric determination of the peracid present.[3] The checkers found that peracetic acid of a lower concentration (27.5%) may also be used without a decrease in yield. The product was found to be sufficiently pure, after only one recrystallization from 60 ml. of petroleum ether (b.p. 40–60°) and cooling overnight to −18°, to be used in the next step.

2. Alternatively *m*-chloroperbenzoic acid (3-chlorobenzenecarboperoxoic acid) may be used.[4]

3. The yields of the desired product decrease substantially if the temperature exceeds 20°.

4. The reported m.p. is 64°.[5] Gas chromatographic analysis using a 1-m. column containing 20% Reoplex 400 on Diatoport S 6080, operated at 160°, indicates the purity of the product to be ∼98%.

5. The rate of bromination of 11-oxatricyclo[4.4.1.0^{1,6}]undeca-3,8-diene is markedly higher in ether than in dichloromethane or chloroform. The former solvent thus permits the reaction to be carried out at relatively low temperatures.

6. Bromine was freshly distilled from phosphorus pentoxide.

7. The addition of bromine should not be started before the solution has cooled to approximately −70°.

8. The tetrabromide apparently consists of a mixture of stereo-isomers. After several recrystallizations from chloroform–petroleum ether (60–90°) the major isomer, m.p. 160–162°, is obtained.

9. As reported by Shani and Sondheimer,[4,6] the dehydrohalogenation of the tetrabromide by means of potassium hydroxide in ethanol at 50–55° affords a mixture, which is readily separated by chromatography on alumina, of 1,6-oxido[10]annulene and the isomeric 1-benzoxepin. The latter compound is also formed during chromatography of 1,6-oxido[10]annulene on silica gel.[7]

10. In order to avoid acid-catalyzed rearrangements of 1,6-oxido-[10]annulene it is recommended that the distillation flask be treated with a base before use.

11. The reported m.p. is 52–53°.[7] The purity of the product is greater than 99% as established by thin layer chromatography, using plates prepared with Silica Gel Si F, obtained from Riedel–De Haen AG, 3016 Seelze, West Germany. If 1,6-oxido[10]annulene is to be used for spectroscopic investigations, care should be taken that its potential contaminants, such as naphthalene, 1-bromonaphthalene, α-naphthol (1-naphthalenol), and 1-benzoxepin, are absent, as checked by thin layer chromatography. The checkers could not obtain the reported yield of 24.5–25.6 g. (68–71%). In experiments where the ether was replaced with tetrahydrofuran or dioxane, the yield given by the submitters could also not be obtained.

12. The spectral properties of the product are as follows; proton magnetic resonance (chloroform-d) δ, multiplicity, assignment: 7.25–7.75 (multiplet, AA′BB′, aromatic H); ultraviolet (95% ethanol) nm. max. (ϵ): 255 (74,000), 299 (6900), 393 (240) complex band; mass

spectrum (250°, 70 e.v.) m/e (relative intensity >10%): 144 (M, 43), 116 (40), 115 (100), 89 (15), 63 (15), 51 (10), 39 (11).

3. Discussion

The preparation of 1,6-oxido[10]annulene, described simultaneously by Sondheimer and Shani[4,6] and by Vogel, Biskup, Pretzer, and Böll,[7] is illustrative of the rather general synthesis of aromatic 1,6-bridged [10]annulenes starting from 1,4,5,8-tetrahydronaphthalene. Apart from the present compound, the following bridged [10]annulenes have thus far been obtained by this approach: 1,6-methano[10]-annulene,[2,8,9] the 11,11-dihalo-1,6-methano[10]annulenes,[9,10] and 1,6-imino[10]annulene.[11]

The first step in this preparation, the epoxidation of 1,4,5,8-tetra-hydronaphthalene, exemplifies the well-known selectivity exerted by peracids in their reaction with alkenes possessing double bonds that differ in the degree of alkyl substitution.[12] As regards the method of aromatization employed in the conversion of 11-oxatricyclo[4.4.1.0^{1,6}]-undeca-3,8-diene to 1,6-oxido[10]annulene, the two-step bromination–dehydrobromination sequence is given preference to the one-step DDQ-dehydrogenation, which was advantageously applied in the synthesis of 1,6-methano[10]annulene,[2,9] since it affords the product in higher yield and purity.

1,6-Oxido[10]annulene closely resembles 1,6-methano[10]annulene in many of its spectral properties, particularly in its proton magnetic resonance, ultraviolet, infrared, and electron spin resonance spectra,[13] but is chemically less versatile than the hydrocarbon analog due to its relatively high sensitivity toward proton and Lewis acids.

1. Institut für Organische Chemie der Universität Köln, 5 Köln, Zülpicher Strasse 47, West Germany.
2. E. Vogel, W. Klug, and A. Breuer, *Org. Syn.*, **54**, 11 (1974).
3. F. P. Greenspan and D. G. MacKellar, *Anal. Chem.*, **20**, 1061 (1948).
4. A. Shani and F. Sondheimer, *J. Amer. Chem. Soc.*, **89**, 6310 (1967).
5. W. Hückel and H. Schlee, *Chem. Ber.*, **88**, 346 (1955).
6. F. Sondheimer and A. Shani, *J. Amer. Chem. Soc.*, **86**, 3168 (1964).
7. E. Vogel, M. Biskup, W. Pretzer, and W. A. Böll, *Angew. Chem.*, **76**, 785 (1964); *Angew. Chem. Int. Ed. Engl.*, **3**, 642 (1964).
8. E. Vogel and H. D. Roth, *Angew. Chem.*, **76**, 145 (1964); *Angew. Chem. Int. Ed. Engl.*, **3**, 228 (1964).
9. P. H. Nelson and K. G. Untch, *Tetrahedron Lett.*, 4475 (1969).

10. V. Rautenstrauch, H.-J. Scholl, and E. Vogel, *Angew. Chem.*, **80**, 278 (1968); *Angew. Chem. Int. Ed. Engl.*, **7**, 288 (1968).
11. E. Vogel, W. Pretzer, and W. A. Böll, *Tetrahedron Lett.*, 3613 (1965).
12. D. Swern, "Organic Peroxides," Vol. 2, Wiley-Interscience, New York, 1971, p. 452.
13. F. Gerson, E. Heilbronner, W. A. Böll, and E. Vogel, *Helv. Chim. Acta*, **48**, 1494 (1965).

PHASE-TRANSFER ALKYLATION OF NITRILES:
α-PHENYLBUTYRONITRILE

(Butanenitrile, 2-phenyl)

$$C_6H_5CH_2CN + C_2H_5Br \xrightarrow[\text{[(C}_2\text{H}_5)_3\text{NCH}_2\text{C}_6\text{H}_5]^+\text{Cl}^-]{\text{NaOH}} C_6H_5\underset{\underset{C_2H_5}{|}}{C}HCN$$

Submitted by M. MAKOSZA and A. JOŃCZYK[1]
Checked by HAROLD W. WAGNER and RICHARD E. BENSON

1. Procedure

A 3-l., four-necked, round-bottomed flask is equipped with a mechanical stirrer, a dropping funnel, a thermometer, and an efficient reflux condenser. To the flask are added 540 ml. of aqueous 50% sodium hydroxide, 257 g. (253 ml., 2.2 moles) of phenylacetonitrile (Note 1), and 5.0 g. (0.022 mole) of benzyltriethylammonium chloride (Note 2). Stirring is begun, and 218 g. (150 ml., 2 moles) of bromoethane (Note 3) is added dropwise over a period of approximately 100 minutes at 28–35°. If necessary, the flask may be cooled by means of a cold-water bath to keep the temperature of the mixture at 28–35°. After the addition of bromoethane is complete, stirring is continued for 2 hours, and then the temperature is increased to 40° for an additional 30 minutes. The reaction mixture is cooled to 25°, 21.2 g. (20.3 ml., 0.2 mole) of benzaldehyde (Note 4) is added, and stirring is continued for 1 hour. The flask is immersed in a cold-water bath, and 750 ml. of water and 100 ml. of benzene are added. The layers are separated, and the aqueous phase is extracted with 200 ml. of benzene. The organic layers are combined and washed successively with 200 ml. of water, 200 ml. of aqueous dilute hydrochloric acid (Note 5), and 200 ml. of water. The organic layer is dried over anhydrous magnesium sulfate, and the solvent is removed by distillation under reduced pressure. The product is distilled through a Vigreux column to give 225–242 g. (78–84%) of α-phenylbutyronitrile, b.p. 102–104° (7 mm.), n^{25} D 1.5065–1.5066 (Notes 6–8).

2. Notes

1. The checkers used phenylacetonitrile obtained from Aldrich Chemical Company, Inc. The product was distilled before use. It may also be purified according to the directions given in *Org. Syn.*, Coll. Vol. **1**, 108 (1948).

2. Benzyltriethylammonium chloride is available from Fisher Scientific Company. The preparation of this reagent is described in *Org. Syn.*, **55**, 97 (1975).

3. Bromoethane (available from Fisher Scientific Company) was distilled before use.

4. Benzaldehyde (available from Fisher Scientific Company) was distilled before use. It is added at this point to convert any unreacted phenylacetonitrile to the high-boiling α-phenylcinnamonitrile [α-(phenylmethylene)benzeneacetonitrile] (Note 7).

5. The acid solution was prepared by adding 1 volume of acid to 5 volumes of water.

6. The checkers obtained a forerun of 7–12 g. of product having n^{25} D 1.5065–1.5066.

7. The α-phenylcinnamonitrile (Note 4) present in the distillation flask can be recovered. The residue is broken up with 75 ml. of methanol, the mixture stirred and cooled, and the product recovered by filtration. Recrystallization from methanol gives 17–20 g. of crystalline material, m.p. 86–88°. The proton magnetic resonance spectrum (chloroform-d) shows complex multiplets at δ 7.20–8.00.

8. Gas chromatographic analysis on a column packed with silicone gum nitrile on acid-washed Gas Chrome Red, 80–100 mesh and heated at 150°, shows that the product is about 97% pure. The material has the following spectral properties; infrared (neat) cm.$^{-1}$: 2250 (C≡N), 1610, 1590 shoulder and 1500 (aromatic C=C), 1385 (C—CH$_3$), and 760 and 697 (monosubstituted aromatic); proton magnetic resonance (carbon tetrachloride) δ, multiplicity, number of protons, assignment, coupling constant J in Hz.: 0.99 (triplet, 3, CH_3, J = 7), 1.82 (multiplet, 2, CH_2), 3.70 (triplet, 1, CH, J = 7), and 7.32 (singlet, 5, C$_6H_5$).

3. Discussion

This reaction is illustrative of a general procedure for the alkylation of active methylene functions in the presence of concentrated aqueous alkali catalyzed by tetraalkylammonium salts. This catalytic method has been used to alkylate arylacetonitriles with monohaloalkanes,[2] dihaloalkanes,[3] α-chloroethers,[4] chloronitriles,[5] haloacetic acid esters,[6] and halonitro aromatic compounds.[7] It has also been used to alkylate ketones,[8] 1*H*-indene,[9] 9*H*-fluorene,[10] and the Reissert compound.[11] The reaction is inhibited by alcohols and by iodide ion.[2]

Methods for the alkylation of nitriles have been reviewed.[12] These procedures, as well as those applied to other active methylenes, generally involve the use of dangerous and expensive condensing agents (sodium amide, metal hydrides, triphenylmethide, potassium *tert*-butoxide, etc.) and strictly anhydrous organic solvents (ether, benzene, *N,N*-dimethylformamide, dimethylsulfoxide, etc.) or liquid ammonia. The catalytic method is much simpler than these others and generally gives good yields of purer products. Because of its high selectivity[13] it is particularly adapted to the synthesis of pure monoalkyl derivatives of phenylacetonitrile which have also been obtained by alkylation of ethyl α-cyanophenylacetate[14] or α-cyanophenylacetic acid,[15] followed by elimination of the carbethoxyl or carboxyl groups.

The catalytic conditions (aqueous concentrated sodium hydroxide and tetraalkylammonium catalyst) are very useful in generating dihalocarbenes from the corresponding haloforms. Dichlorocarbene thus generated reacts with alkenes to give high yields of dichlorocyclopropane derivatives,[16] even in cases where other methods have failed,[17] and with some hydrocarbons to yield dicholromethyl derivatives.[18] Similar conditions are suited for the formation and reactions of dibromocarbene,[19] bromofluoro- and chlorofluorocarbene,[20] and chlorothiophenoxy carbene,[21] as well as the Michael addition of trichloromethyl carbanion to unsaturated nitriles, esters, and sulfones.[22]

This method exemplifies a broad class of processes that proceed *via* transfer of reacting species between two liquid phases. Such processes may require a catalyst that can combine with species present in one phase and effect their transfer in this form to the second phase where the main reaction occurs. Starks[23] has termed such a process "phase-transfer catalysis" and has demonstrated its utility in reactions involving inorganic anions. For example, he has shown that the rates

TABLE I
ALKYLATIONS IN AQUEOUS MEDIUM

Compound	Alkylation Agent	Product	(%) Yield	Reference
$C_6H_5CH_2CN$	$(C_6H_5)_2CHCl$	$(C_6H_5)_2CHCH(C_6H_5)CN$	94	2
$C_6H_5CH_2CN$	$Br(CH_2)_4Br$		88	3
$C_6H_5CH(C_2H_5)CN$	$C_6H_5CH_2Cl$	$C_6H_5CH_2C(C_6H_5)CN$	94	2
$(C_6H_5)_2CHCN$	$BrCH_2CH_2Br$	$(C_6H_5)_2C(CH_2CH_2Br)CN$	91	3
$C_6H_5CH(CH_3)CN$	$ClCH_2OCH_3$	$C_6H_5C(CH_2OCH_3)CN$	68	4
$C_6H_5CH(CH_3)CN$	$4\text{-}ClC_6H_4NO_2$	$C_6H_5C(4\text{-}NO_2C_6H_4)CN$	82	11
$C_6H_5CH(C_2H_5)CN$	$ClCH_2COOC_4H_9\text{-}tert$	$C_6H_5C(CH_2COOC_4H_9\text{-}tert)CN$	77	6
$C_6H_5CH_2COCH_3$	$BrCH_2CH_2Br$		54	8
	$Br(CH_2)_4Br$		64	9

94

of some displacement, oxidation, and hydrolysis reactions conducted in two-phase systems are dramatically enhanced by the presence of ammonium and phosphonium salts. However in reactions involving weakly active methylenes, the catalyst seems to be more than a simple transfer agent; it is necessary for carbanion formation.

The versatility of this method for the alkylation of compounds containing active methylene groups is illustrated by Table I. Review articles have recently appeared,[24] and the application to the Hofmann carbylamine reaction is described in the following procedure in this volume, p. 96.

1. Institute of Organic Chemistry and Technology, Technical University (Politechnika), Warsaw, Poland.
2. M. Makosza and B. Serafin, *Rocz. Chem.*, **39**, 1223, 1401, 1595, 1805 (1965) [*C.A.*, **64**, 12595h, 17474g, 17475c, 17475g (1966)].
3. M. Makosza and B. Serafin, *Rocz. Chem.*, **40**, 1647, 1839, (1966) [*C.A.*, **66**, 94792x, 115435a (1967)].
4. M. Makosza, B. Serafinowa, and M. Jawdosiuk, *Rocz. Chem.*, **41**, 1037 (1967) [*C.A.*, **68**, 39313h (1968)].
5. J. Lange and M. Makosza, *Rocz. Chem.*, **41**, 1303 (1967) [*C.A.*, **68**, 29374q (1968)].
6. M. Makosza, *Rocz. Chem.*, **43**, 79 (1969) [*C.A.*, **70**, 114776h (1969)].
7. M. Makosza, *Tetrahedron Lett.*, 673 (1969); M. Makosza and M. Ludwikow, *Bull. Acad. Pol. Sci., Ser. Sci. Chim.*, **19**, 231 (1971) [*C.A.*, **75**, 48646r (1971)].
8. A. Jończyk, B. Serafin, and M. Makosza, *Tetrahedron Lett.*, 1351 (1971).
9. M. Makosza, *Tetrahedron Lett.*, 4621 (1966).
10. M. Makosza, *Bull. Acad. Pol. Sci., Ser. Sci. Chim.*, **15**, 165 (1967) [*C.A.*, **67**, 64085x (1967)].
11. M. Makosza, *Tetrahedron Lett.*, 677 (1969).
12. A. C. Cope, H. L. Holmes, and H. O. House, *Org. React.*, **9**, 107 (1957); M. Makosza, *Wiad. Chem.*, **21**, 1 (1967) [*C.A.*, **67**, 53161t (1967)]; M. Makosza, *Wiad. Chem.*, **23**, 35, 759 (1969) [*C.A.*, **70**, 96065u (1969); *C.A.*, **72**, 110907v (1970)].
13. M. Makosza, *Tetrahedron*, **24**, 175 (1968).
14. R. Delaby, P. Reynaud, and F. Lily, *Bull. Soc. Chim. Fr.*, 864 (1960).
15. E. M. Kaiser and C. R. Hauser, *J. Org. Chem.*, **31**, 3873 (1966).
16. M. Makosza and M. Wawrzyniewicz, *Tetrahedron Lett.*, 4659 (1969).
17. E. V. Dehmlow and J. Schönefeld, *Justus Liebigs Ann. Chem.*, **744**, 42 (1971).
18. I. Tabushi, Z. Yoshida, and N. Takahashi, *J. Amer. Chem. Soc.*, **92**, 6670 (1970); E. V. Dehmlow, *Tetrahedron*, **27**, 4071 (1971).
19. M. Makosza and M. Fedorynski, *Bull. Acad. Pol. Sci., Ser. Sci. Chim.*, **19**, 105 (1971) [*C.A.*, **75**, 19745s (1971)].
20. P. Weyerstahl, G. Blume, and C. Müller, *Tetrahedron Lett.*, 3869 (1971).
21. M. Makosza and E. Bialecka, *Tetrahedron Lett.*, 1517 (1971).
22. M. Makosza and I. Gajos, *Bull. Acad. Pol. Sci., Ser. Sci. Chim.*, **20**, 33 (1972) [*C.A.*, **76**, 153179j (1972)].
23. C. M. Starks, *J. Amer. Chem. Soc.*, **93**, 195 (1971).
24. J. Dock, **Synthesis**, 441 (1973); E. V. Dehmlow, *Angew. Chem. Int. Ed. Engl.*, **13**, 170 (1974).

PHASE-TRANSFER
HOFMANN CARBYLAMINE REACTION: *tert*-BUTYL ISOCYANIDE
(2-Methylpropane, 2-isocyano)

$$(CH_3)_3CNH_2 + CHCl_3 + 3\ NaOH \xrightarrow[\text{H}_2\text{O, dichloromethane, reflux}]{[(C_2H_5)_3NCH_2C_6H_5]^+Cl^-}$$

$$(CH_3)_3CN{=}C + 3\ NaCl$$

Submitted by George W. Gokel, Ronald P. Widera,
and William P. Weber[1]
Checked by F. A. Souto-Bachiller, S. Masamune,
Charles J. Talkowski, and William A. Sheppard

1. Procedure

Caution! This preparation should be conducted in an efficient hood because of the evolution of carbon monoxide and the obnoxious odor of the isocyanide.[2]

A 2-l., round-bottomed flask equipped with a magnetic stirring bar, a reflux condenser, and a pressure-equalizing dropping funnel is charged with 300 ml. of water. Stirring is commenced and 300 g. (7.50 moles) of sodium hydroxide is added in portions in order to maintain smooth stirring (Note 1). The funnel is charged with a mixture containing 141.5 g. (208 ml., 1.93 moles) of *tert*-butylamine, 117.5 g. (80 ml., 0.98 mole) of chloroform (Note 2), and 2 g. of benzyltriethyl-ammonium chloride (Note 3) in 300 ml. of dichloromethane. The mixture is added dropwise to the stirred, warm (*ca.* 45°) solution over a 30-minute period. The reaction mixture begins to reflux immediately after initiation of the addition (Note 4) and subsides within 2 hours; stirring is continued for an additional hour (Note 5). The reaction mixture is diluted with 800 ml. of ice and water, and the organic layer is separated and retained. The aqueous layer is extracted with 100 ml. of dichloromethane, and the dichloromethane solution is added to the initial organic phase. The resulting solution is successively washed with 100 ml. of water, 100 ml. of aqueous 5% sodium chloride, and is dried over anhydrous magnesium sulfate.

The drying agent is removed by filtration, and the filtrate is distilled under nitrogen through a spinning band column (Notes 6 and 7). The fraction, boiling at 92–93° (725 mm.), is collected to yield 54.2-60.0 g. (66–73%, based on chloroform) of *tert*-butyl isocyanide (Notes 8 and 9).

2. Notes

1. Efficient stirring is required. A solution of 225 g. (5.6 moles) of sodium hydroxide in 225 ml. of water can be added to the stirred mixture of the organic substrates in dichloromethane if a more efficient mechanical stirrer is used. In the original procedure, the submitters noted an induction period of about 20 minutes which was stated to vary somewhat with the stirring rate, stirring-bar size, and relative amount of phase-transfer catalyst. Three moles of base are required for the reaction: one to generate the carbene and two to react with the additional two moles of hydrochloric acid lost by the amine–carbene adduct in the isonitrile formation step. If less base is used, the excess hydrochloric acid reacts with the isonitrile by α-addition, and the yield is substantially reduced.

2. Chloroform, commercially available, normally contains 0.75% ethanol and was used as supplied.

3. Benzyltriethylammonium chloride is available from Eastman Organic Chemicals. The checkers prepared the salt in a state of high purity by a modification of a reported procedure.[3] A solution of 33.7 g. (0.33 mole) of triethylamine and 50.0 g. (0.40 mole) of benzyl chloride (both from Eastman Organic Chemicals) in 60 ml. of absolute ethanol was refluxed for 64 hours. The solution was cooled to room temperature and 300 ml. of ether was added. The precipitated ammonium salt was removed by filtration, redissolved in the minimum amount of hot acetone, and reprecipitated with ether.

4. The volatilities of both *tert*-butylamine and dichloromethane necessitate the use of an efficient condenser as a precaution, although the rate of reflux is generally not vigorous. In preparations where higher boiling amines are used, this precaution is less critical.

5. The submitters noted that a longer stirring period did not seem to affect the yield appreciably.

6. The bulk of the residual *tert*-butylamine is recovered.

7. A 60-cm. annular Teflon® spinning band distillation column is recommended to achieve clean separation of solvent and unreacted reagent from product rather than a column packed with glass helices. For higher boiling isocyanides, separation of solvent and unreacted reagents may be effected by the use of a rotary evaporator, although the thermal instability of the isocyanides should be taken into consideration.

8. Yields higher than about 70% for any of these isonitrile preparations generally indicate incomplete fractionation. The purity of the product may be conveniently checked by proton magnetic resonance spectroscopy. The characteristic 1:1:1 triplet for *tert*-butyl isocyanide appears at δ 1.45 (chloroform-*d*). A small upfield peak usually indicates the presence of unreacted amine. Other common contaminants are dichloromethane and chloroform. The purity may be determined more accurately by gas chromatographic analysis on a 230 cm. by 0.6 cm. column packed with 10%SE30 on Chromosorb G, 60–80 mesh, at 80°.

9. Glassware may be freed from the isocyanide odor by rinsing with a 1:10 mixture of concentrated hydrochloric acid and methanol.

3. Discussion

The present method utilizes dichlorocarbene generated by the phase-transfer method of Makosza[4] and Starks.[5] The submitters have routinely realized yields of pure distilled isocyanides in excess of 40%.[6] With less sterically hindered primary amines a 1:1 ratio of amine to chloroform gives satisfactory results. Furthermore, by modifying the procedure, methyl and ethyl isocyanides may be prepared directly from the corresponding aqueous amine solutions and bromoform.[7] These results are summarized in Table I.

TABLE I

PREPARATION OF ISOCYANIDES (RN=C) BY THE CARBYLAMINE REACTION[a]

R	$CHCl_3 + NaOH \xrightarrow{PTC} [:CCl_2] \xrightarrow{RNH_2} RN=C$ Yield (%)	b.p. of RN=C
$CH_3(CH_2)_2CH_2-$	60	40–42° (11 mm.)
$C_6H_5CH_2-$	45	92–93° (11 mm.)
$CH_3(CH_2)_{10}CH_2-$	41	115–118° (0.1 mm.)
$c-C_6H_{11}-$	48	67–72° (13 mm.)
C_6H_5-	57	50–52° (11 mm.)
CH_3-[b]	24	59–60° (760 mm.)
CH_3CH_2-[b]	47	78–79° (760 mm.)

[a] Prepared by the phase-transfer method using chloroform and aqueous sodium hydroxide with the corresponding amines.[6]

[b] Bromoform substituted for chloroform for ease of fractionation.[7]

Various synthetic routes to isocyanides have been reported since their identification over 100 years ago.[8] Until now, the useful synthetic procedures all required a dehydration reaction.[9-11] Although the carbylamine reaction involving the dichlorocarbene intermediate is one of the early methods,[8] it had not been preparatively useful until the innovation of phase-transfer catalysis (PTC).[4,5]

The phase-transfer catalysis method has also been utilized effectively for addition of dichlorocarbene to olefins,[4] as well as for substitution and elimination reactions, oxidations, and reductions.[12] The preceding procedure in this volume is another example.[13]

1. Department of Chemistry, University of Southern California, University Park, Los Angeles, California 90007. This work was supported in part by a grant from the National Science Foundation, grant number GP-40331X.
2. Many isocyanides are reported to exhibit no appreciable toxicity to mammals. See J. A. Green, II and P. T. Hoffmann in "Isonitrile Chemistry," I. Ugi (Ed.), Academic Press, New York, 1971, p. 2. However, since certain isocyanides are highly toxic (e.g., 1,4-diisocyanobutane), the checkers recommend that all isocyanides be handled with due caution.
3. R. A. Moss and W. L. Sunshine, *J. Org. Chem.*, **35**, 3581 (1970).
4. M. Makosza and M. Wawrzyniewicz, *Tetrahedron Lett.*, 4659 (1969).
5. C. M. Starks, *J. Amer. Chem. Soc.*, **93**, 195 (1971).
6. W. P. Weber and G. W. Gokel, *Tetrahedron Lett.*, 1637 (1972).
7. W. P. Weber, G. W. Gokel, and I. Ugi, *Angew. Chem. Int. Ed. Engl.*, **11**, 530 (1972).
8. For a review, see P. T. Hoffmann, G. Gokel, D. Marquarding, and I. Ugi in "Isonitrile Chemistry," I. Ugi (ed.), Academic Press, New York, 1971, Chapter II.
9. I. Ugi, R. Meyr, M. Lipinski, F. Bodesheim, and F. Rosendahl, *Org. Syn.*, Coll. Vol. **5**, 300 (1973); I. Ugi and R. Meyr, *Org. Syn.*, Coll. Vol. **5**, 1060 (1973).
10. G. E. Niznik, W. H. Morrison, III, and H. M. Walborsky, *Org. Syn.*, **51**, 31 (1971).
11. R. E. Schuster, J. E. Scott, and J. Casanova, Jr., *Org. Syn.*, Coll. Vol. **5**, 772 (1973).
12. J. Dockx, *Synthesis*, 441 (1973); E. V. Dehmlow, *Angew. Chem. Int. Ed. Engl.*, **13**, 170 (1974); A. W. Herriott and D. Picker, *Tetrahedron Lett.*, 1511 (1974).
13. M. Makosza and A. Jónczyk, *Org. Syn.*, **55**, 91 (1975).

2-PHENYL-2-VINYLBUTYRONITRILE

(3-Butenenitrile, 2-ethyl-2-phenyl)

$$C_6H_5CHCN + HC\equiv CH \xrightarrow[\text{dimethyl sulfoxide, 60° to 70°}]{[(C_2H_5)_3NCH_2C_6H_5]^+Cl^-, KOH} C_6H_5CCN$$

with C_2H_5 below the left carbon, and $CH\!=\!CH_2$ above and C_2H_5 below the right carbon.

Submitted by M. Makosza, J. Czyzewski, and M. Jawdosiuk[1]
Checked by John C. Sauer and Richard E. Benson

174843

1. Procedure

A 1-l., four-necked, round-bottomed flask is equipped with a sealed mechanical stirrer, a thermometer, and a gas-inlet tube. To the flask are added 145 g. (1 mole) of 2-phenylbutyronitrile (2-phenylbutanenitrile) (Note 1), 2.3 g. (0.01 mole) of benzyltriethylammonium chloride (N,N,N-triethylbenzenemethanaminium chloride) (Note 2), and 50 ml. of dimethyl sulfoxide (Note 3). The gas-inlet tube is adjusted to extend below the surface of the liquid, and a gas-exit tube is attached to the flask. A slow stream of acetylene (ethyne) (Note 4) is passed through the gas-inlet tube into the flask in order to remove the air. After 5 minutes, 56 g. of finely powdered potassium hydroxide is added and stirring is begun. Acetylene is introduced at the rate of 15–20 l./hour. An exothermic reaction occurs; the temperature rises to 70–80° and is held in this range by means of a cold-water bath (Note 5). After 40–60 minutes, a warm bath is required to maintain the temperature of the reaction mixture at 60–70°, and stirring is continued for an additional 20–30 minutes (Note 6). The mixture is cooled to room temperature, the inlet tube is replaced by a pressure-equalizing dropping funnel, and 500 ml. of water is added slowly (Note 7). The resultant dark brown mixture is transferred to a separatory funnel and washed two times with 200-ml. portions of benzene. The benzene layers are combined and washed successively with 200 ml. of water, 100 ml. of aqueous 10% hydrochloric acid, and 200 ml. of water. The organic layer is dried over anhydrous magnesium sulfate, and the benzene is removed by distillation at reduced pressure. The residual oil is distilled through a short Vigreux column to give 125–135 g. of crude product, b.p. 115–125° (13 mm.). Redistillation of this product through a Vigreux column gives 101–107 g. (59–63%) of colorless 2-phenyl-2-vinylbutyronitrile, b.p. 110° (8 mm.), n^{25} D 1.5157 (Note 8).

2. Notes

1. The preparation of 2-phenylbutyronitrile is described in *Org. Syn.*, **55**, 91 (1975).

2. The checkers used the product available from Aldrich Chemical Company, Inc. The preparation of this reagent is described in the preceding procedure.

3. The checkers used the product available from Fisher Scientific Company.

4. The checkers used acetylene available from Matheson Gas Products. The gas was purified by passing it through concentrated sulfuric acid and then through a tower filled with potassium hydroxide pellets. The gas was then passed into a 1-l. safety flask which was connected to the gas inlet tube by means of rubber tubing. The checkers used a rotameter that was calibrated with air to determine the flow rate of acetylene.

5. The checkers attempted to keep the temperature at 65–70°.

6. The reaction may be monitored by gas chromatography. The submitters used a 2-m. column containing silicone oil on diatomite support (190°).

7. The water should be added slowly, since the mixture is saturated with acetylene and the gas may be evolved vigorously.

8. The checkers found the product to be at least 95% pure on a gas chromatographic column containing 10% silicone 200 on nonacid washed Chromosorb W operated at 125°. The spectral properties of the product are as follows; infrared (neat) cm.$^{-1}$: 1639, 1000, 930 (CH= CH_2); proton magnetic resonance (neat) δ, multiplicity, number of protons, assignment, coupling constant J in Hz.: 0.92 (triplet, 3, CH_3, $J = 7$), 1.87 (quartet, 2, CH_2, $J = 7$), 5.00–6.20 (multiplet, 3, $CH=CH_2$), 7.17–7.57 (multiplet, 5, C_6H_5).

3. Discussion

This procedure, which involves the addition of an anion derived from a nitrile to an unactivated acetylenic bond under rather mild conditions, is a convenient general method for the synthesis of α-vinyl-nitriles (see Table I). The reaction proceeds smoothly in either dimethyl sulfoxide or hexamethylphosphoric triamide (*see p. 103 for warning*) as solvent with a tetraalkylammonium salt as catalyst. The products thus prepared are obtained in yields higher[2] than those obtained under conventional conditions, which generally require higher temperatures and elevated pressures.[3,4]

1. Institute of Organic Chemistry and Technology, Technical University (Politechnika), Warsaw, Poland.
2. M. Makosza, Pol. Patent **55113** (1968) [*C.A.*, **70**, 106006s (1969)]; M. Makosza, *Tetrahedron Lett.*, 5489 (1966); M. Makosza and M. Jawdosiuk, *Bull. Acad. Pol. Sci.*, *Ser. Sci. Chim.*, **16**, 589 (1968) [*C.A.*, **71**, 30193y (1969)].
3. P. P. Karpukhin and A. I. Levchenko, *Zh. Prikl. Khim. Leningrad*, **32**, 1354 (1959) [*C.A.*, **54**, 450c (1960)]; P. P. Karpukhin, A. I. Levchenko, and E. V. Dudko, *Zh. Prikl. Khim. Leningrad*, **34**, 1117 (1961) [*C.A.*, **55**, 22259f (1961)].
4. M. Seefelder, *Justus Liebigs Ann. Chem.*, **652**, 107 (1962).

TABLE I

α-Vinylnitriles Derived from Acetylenes

Nitrile	Acetylene	Product	b.p.	Yield (%)
$C_6H_5CH(C_5H_{11})CN$	$HC{\equiv}CH$	$C_6H_5C(C_5H_{11})CN$ $\vert$ $CH{=}CH_2$	139° (8 mm.)	88
$C_6H_5CH(2{\text -}C_3H_7)CN$	$HC{\equiv}CC_6H_5$	$C_6H_5C(2{\text -}C_3H_7)CN^a$ $\vert$ $CH{=}CHC_6H_5$	145° (0.8 mm.)	83
$(C_6H_5)_2CHCH(C_6H_5)CN$	$HC{\equiv}CSC_4H_9$	$(C_6H_5)_2CHC(C_6H_5)CN^b$ $\vert$ $CH{=}CHSC_4H_9$	113°c	96
$(C_6H_5)_2CHCN$	$HC{\equiv}COC_2H_5$	$(C_6H_5)_2C{-}C{=}CH_2$ $\vert \quad \vert$ $CN \; OC_2H_5$	146° (0.6 mm.)	77

a Product is a 2:1 mixture of (Z) and (E) isomers; (Z) isomer, m.p. 62°.
b Only the (Z) isomer is obtained.
c Melting point.

PREPARATION OF ALKENES BY REACTION OF LITHIUM DIPROPENYLCUPRATES WITH ALKYL HALIDES: (E)-2-UNDECENE

Submitted by Gerard Linstrumelle,[1,2] Jeanne K. Krieger,[1] and George M. Whitesides[1]
Checked by Gordon S. Bates and S. Masamune

1. Procedure

*Caution! Lithium dispersion is highly reactive. Although it may be handled safely in air when covered with mineral oil, it will spark and ignite organic vapors on contact with water. When clean, it is pyrophoric and may occasionally ignite spontaneously on contact with air. Manipulation of lithium dispersion should be carried out behind a safety shield with the protection of rubber gloves. Particular caution should be taken during disposal of the excess lithium remaining after preparation and filtration of the propenyllithium (Note 1). Hexamethylphosphoric triamide (HMPA) vapors have been reported to cause cancer in rats [J. A. Zapp, Jr., Science, **190**, 422 (1975)]. Hence, all operations with HMPA should be performed in a good hood, and care should be taken to keep liquid HMPA off the skin.*

A. *(E)-1-Propenyllithium.* A dry (Note 2), 500-ml., three-necked, round-bottomed flask equipped with a Teflon®-covered magnetic stirring bar, a 200-ml. pressure-equalizing dropping funnel, an efficient reflux condenser, and an immersion thermometer is capped with serum stoppers (Note 3) and flushed with argon (Note 4). Lithium dispersion (Note 5), [22.4 g. of a 50% *w/w* suspension in Amsco,

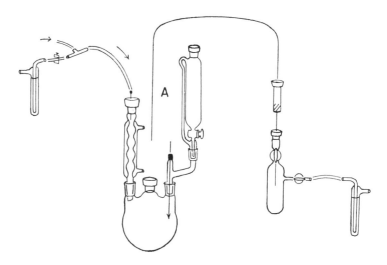

11.2 g. (1.6 g.-atoms) of lithium] (Note 6), is rapidly weighed in air and transferred to the flask. The apparatus is flushed again with argon for approximately 10 minutes, and then maintained under a static pressure of the inert gas (Note 7). The lithium is washed three times by transferring approximately 60-ml. portions of anhydrous ether (Note 8) into the flask through a No-Air stopper by forced siphon through a stainless steel cannula (Note 9), stirring the resulting suspension of lithium briefly, allowing the suspension to separate, and finally withdrawing the major part of the ether by forced siphon through a cannula. After separating the last washings, 250 ml. of anhydrous ether is transferred into the flask (see diagram), and a solution of 35 g. (0.46 mole) of (E)-1-chloropropene (Note 10) in 120 ml. of anhydrous ether is transferred into the dropping funnel through a cannula. The reaction mixture is cooled in an acetone bath maintained at approximately −10° by periodic addition of dry ice, and the surface of the lithium cleaned by injecting approximately 150 μl. of 1,2-dibromoethane through the serum stopper into the vigorously stirred lithium suspension (Note 11). The chloropropene solution is added to the lithium suspension over a 2-hour period, while the internal temperature of the reaction mixture is maintained at 10° ± 5°. It is important to control the temperature in this range throughout the addition (Note 12). A precipitate begins to form. The reaction mixture is allowed to stir for 30 minutes at room temperature, and stirring is stopped. The precipitate is allowed to settle, and the solution is transferred by a large-gauge cannula through a glass wool pad into a storage bottle in the manner illustrated in the diagram (Note 13). The precipitate is washed once with 20 ml. of anhydrous ether, and this solution is transferred to the storage bottle through the filter. Gilman titration[3] of the resulting solution using aqueous 0.10N hydrochloric acid and 1,2-dibromoethane typically indicates that the concentration of total base is 1.29N and the concentration of residual base is 0.25N (Note 14). The reagent can be stored at $-20°$ for more than one week without isomerization or attack on ether, but should be titrated routinely before use.

B. (E)-2-*Undecene.* A dry (Note 2), 500-ml., three-necked, round-bottomed flask equipped with a thermometer, a 200-ml. pressure-equalizing dropping funnel capped with a No-Air stopper, a gas-inlet tube, and a Teflon®-covered magnetic stirring bar is charged with 12.58 g. (0.066 mole) of copper(I) iodide (Note 15). The apparatus is flushed with argon. Under a static argon pressure of 2–3 cm. of Nujol, 80 ml. of anhydrous ether is added, and the resulting suspension is immersed in an acetone–dry ice bath at $-78°$. Through a cannula 132 ml. of an ethereal 1.04M (0.137 mole) solution of (E)-1-propenyl-lithium is transferred into the dropping funnel and added dropwise to the cold, stirred suspension of copper(I) iodide over 45–50 minutes. The reaction mixture is allowed to warm to $-35°$ over approximately 1 hour (Note 16). At $-35°$, 44 ml. of hexamethylphosphoric triamide (approximately 3.75 equivalents per equivalent of lithium dipropenyl-cuprate) is added slowly to the reaction mixture from the dropping funnel, which has been previously rinsed with 20 ml. of ether (Note 17). A white precipitate forms almost immediately (Note 18). After approximately 10 minutes, 15.84 g. (12 ml., 0.066 mole) of 1-iodooctane is added from the dropping funnel over 10 minutes (Note 19), while the reaction temperature is maintained at $-35°$. The funnel is rinsed with 20 ml. of ether. The resulting mixture is allowed to warm to a temperature between $-5°$ and $0°$ over 45 minutes (Note 20). The entire reaction mixture is then poured into 400 ml. of aqueous saturated ammonium chloride and filtered through fritted glass. The precipitate on the filter is washed with 30 ml. of pentane. The organic layer is separated, and the aqueous phase extracted with four 80-ml. portions of pentane. The combined organic layers are washed once with 40 ml. of aqueous saturated ammonium chloride and once with 20 ml. of aqueous saturated sodium chloride and dried over anhydrous magnesium sulfate. The resulting solution is concentrated by distillation at atmospheric pressure through a 12-cm. Vigreux column, and the residue is distilled at reduced pressure through the same column, yielding 9.2–9.4 g. (90–93%) of (E)-2-undecene, b.p. 88° (17 mm.), n^{25} D 1.4276 (Note 21).

2. Notes

1. Lithium dispersion can be safely destroyed by carefully adding it in small portions to a large excess of technical *tert*-butyl alcohol in a metal pan. If too much lithium is added at one time, the reaction with the *tert*-butyl alcohol can become very vigorous. Under these circumstances, a fire can be avoided by covering the pan with a second, larger metal pan or with an asbestos sheet.

2. The glassware was dried in an oven at 120° for 2 hours and assembled quickly while still hot. It is possible and occasionally desirable in humid climates to dry the apparatus after assembly by heating its accessible parts with a low flame from a Meaker burner while flushing it with a stream of argon, but this operation is normally unnecessary.

3. The serum stoppers used were "sleeve serum stopples," manufactured by the Bittner Corporation, 181 Hudson St., New York, N.Y., 10013; they were obtained from V.W.R. Scientific. The checkers used rubber serum stoppers obtained from Fisher Scientific Company.

4. Argon, available from Matheson Gas Products, was used without drying. *Caution! Nitrogen can NOT be used as a substitute for argon in the preparation of organolithium reagents, since lithium metal reacts rapidly with nitrogen.* Argon and nitrogen can be used interchangeably in subsequent steps. Inert gases are delivered to the reaction vessel through Tygon® or polyethylene tubing, whichever is more convenient.

5. Lithium dispersion containing 1% sodium, so-called "high-sodium" lithium dispersion, suspended in Amsco (a mineral oil) was obtained from Foote Mineral Company. Lithium containing smaller amounts of sodium is not usable in this procedure. This lithium dispersion separates slowly on standing. Before weighing samples of the dispersion, it is stirred briefly to produce a reasonably homogeneous suspension. The checkers used lithium dispersion 50% w/w in cyclohexane obtained from Matheson Coleman and Bell.

6. Experiments carried out using a smaller excess (1.2 g.-atoms) of lithium gave results indistinguishable from those obtained using the

procedure outlined here. Nevertheless, because the lithium suspension in Amsco is inhomogeneous, an excess of lithium is preferred.

7. A slight positive pressure of argon was maintained in the vessel throughout the reaction with an argon line connected to both a bubbler containing Nujol and a hypodermic needle that had been inserted through the serum stopper covering the reflux condenser, as illustrated in the diagram.

8. Mallinckrodt anhydrous diethyl ether was purified by refluxing with calcium hydride under nitrogen for at least 2 hours and distilling under nitrogen immediately before use. For use in the preparation of the propenyllithium, approximately 650 ml. of ether should be collected in a dry 1-l. graduated flask capped with a serum stopper.

9. Cannulae are fabricated on special order from stainless steel hypodermic needle stock by Popper and Sons Company, 290 Park Avenue, South, New York, N.Y., 10010. Approximately 30-cm. long cannulae made from 15 and 20 gauge stock were used in this preparation.

10. Pure (E)-1-chloropropene was obtained by careful distillation of a mixture of (E)- and (Z)-1-chloropropene (available from Columbia Organic Chemicals Company Inc.) using a Nester–Faust Teflon® annular spinning band column [(Z)-1-chloropropene has b.p. 33°; (E)-1-chloropropene has b.p. 37°]. Small quantities of powdered sodium bicarbonate and hydroquinone (1,4-benzenediol) placed in the distillation flask inhibit acid-catalyzed isomerization and polymerization. Gas chromatographic analysis of the material used in these experiments on a 4-m., 15% 1,2,3-tris(2-cyanoethoxy)propane (TCEP) on Chromosorb P column, operated at room temperature, typically indicated that it had isomeric purity >99.9%. (E)-1-Chloropropene is stable for several months at room temperature, but it should be stored in a cool place.

11. Purification of 1,2-dibromoethane was accomplished by passing a small quantity rapidly through a 5-cm. column of alumina. Addition of 1,2-dibromoethane to the lithium dispersion is accompanied by visible evolution of ethylene (ethene).

12. If the internal temperature is too low the reaction stops; therefore the mixture should not be cooled excessively. A buildup of unreacted (E)-1-chloropropene in the reaction flask can result in an uncontrollable exotherm when the reaction temperature again increases.

13. The transfer of filtration were accomplished by inserting the cannula (label A in the diagram) through the serum stopper on the center neck of the reaction flask and to the bottom of the flask to avoid transferring the unreacted lithium dispersion floating at the top of the solution. The pressure used in the forced siphon was controlled with the stopcock on the Nujol bubbler attached to the inlet gas line. A convenient filter was constructed by packing glass wool, previously dried in an oven, into a 20-ml. Luer-lok® syringe barrel, fitted with a 20-cm. 15-gauge needle. The syringe barrel was capped with a serum stopper, flushed with argon, and washed with approximately 30 ml. of anhydrous ether. A large-diameter (15 gauge) cannula should be used to transfer the organolithium reagent solution from the flask to the filter, since smaller gauge cannulae are frequently plugged by solid particles. The storage bottle is usually a Schlenk tube fitted with Teflon® stopcocks. The lithium reagent can also be stored for 2 or 3 days in a flask fitted with a serum stopper, provided that puncture holes through the stopper are sealed with Apiezon Q or some equivalent material.

14. After this titration, the isomeric purity of the organolithium reagent can be assayed by adding approximately 4 ml. of pentane to the dibromoethane–water mixture, separating the organic layer, drying over anhydrous magnesium sulfate, filtering, and analyzing by gas chromatography at room temperature. Less than 1% of unreacted (E)-1-chloropropene was present; the (E)-1-bromopropene was accompanied by approximately 3.5% of (Z)-1-bromopropene.

15. Commercial copper(I) iodide (purchased from Fisher Scientific Company) was purified by the published procedure.[4a] The very pale yellow powder obtained was collected by suction filtration on a fritted glass filter, then triturated on the filter in succession with four 100-ml. portions of distilled water, four 80-ml. portions of reagent-grade acetone, and four 80-ml. portions of anhydrous ether. The copper(I) iodide was powdered, transferred into a round-bottomed flask fitted with a connecting tube with a stopcock, dried overnight under vacuum (0.6 mm.) at room temperature, and heated at 90° under the same vacuum for 4 hours. After cooling, nitrogen was introduced. When copper(I) iodide was needed, it was weighed, powdered, and transferred rapidly in air without special precautions. *Each time the flask is opened to remove a sample of copper(I) iodide, the air introduced should be removed by evacuating the flask under vacuum and refilling it with nitrogen*

through a two-way stopcock several times. If this precaution is observed, the copper(I) iodide can be stored without change for several months; if air is allowed to remain in the flask, the copper(I) iodide deteriorates.[4b]

16. Occasionally a colorless solution was obtained at approximately −50°. In most cases the reaction mixture contained a small quantity of fine suspended black powder (presumably copper metal). Examination of the reaction mixture at this stage by hydrolysis of a sample at −35° and gas chromatographic analysis demonstrated the presence of approximately 10% of 2,4-hexadienes [predominantly the (*E*),(*E*)-isomer].

17. Hexamethylphosphoric triamide, supplied by Fisher Scientific Company, was purified by distillation from sodium (approximately 6 g. per 500 ml.). The boiling point was 69° (1 mm.). In place of hexamethylphosphoric triamide, tetrahydrofuran can be used, and 120 ml. of this solvent was slowly added over 30 minutes through a stainless steel cannula. Mallinckrodt analytical-reagent tetrahydrofuran was purified by distillation from a dark purple solution of disodium benzophenone dianion under nitrogen before use. The checkers used only hexamethylphosphoric triamide (*see p. 103 for warning*).

18. A major part of this precipitate was apparently crystallized hexamethylphosphoric triamide, presumably containing lithium halide.

19. 1-Iodooctane (purchased from Eastman Organic Chemicals) was purified by distillation, b.p. 57° (0.7 mm.), and stored at 0° in the dark.

20. The reaction is complete within 30 minutes if the temperature of the bath is maintained between −30° and −25°.

21. Gas chromatographic analyses for isomeric purity were carried out using a TCEP column (Note 10) at 70°; these analyses typically indicated an isomeric composition of 96% (*E*)- and 4% (*Z*)-2-undecene. Examination of the crude product before distillation showed the presence of approximately 10% 2,4-hexadienes (Note 16). (*E*)-2-Undecene has a shorter retention time than (*Z*)-2-undecene on the column used here. Spectral properties of the (*E*)-isomer are as follows; infrared (neat) cm.$^{-1}$: 3020 weak inflection, 2960 medium, 2935 very strong, 2865 strong, 1560 weak, 1455 medium, 1380 weak, 975 medium-strong; proton magnetic resonance (carbon tetrachloride) δ, multiplicity, number of protons, assignment: 0.9 (multiplet, 3, allylic CH_3), 1.3 (broad singlet, 12, 6 × CH_2), 1.6 (multiplet, 3, CH_3), 1.9 (multiplet, 2, allylic CH_2), 5.35 (multiplet, 2, CH=CH).

3. Discussion

This procedure illustrates the stereospecific preparation of a lithium vinylcuprate and its utilization in the stereospecific synthesis of an (E)-alkene by coupling with an organic halide. Similar results have been obtained with lithium di-(Z)-propenylcuprate, prepared from (Z)-1-bromopropene,[3] and such procedures can be used to prepare alkenes from (Z)- and (E)-vinylic halides.[5] These coupling procedures seem widely applicable in reactions involving primary alkyl halides or tosylates. Using the conditions described above, yields are variable but generally lower when applied to secondary halides (35%), where elimination becomes an important side reaction, and to aryl halides (25%), where metal–halogen exchange may become important.[6,7] No systematic effort has been made to minimize these side reactions. Vinylcuprates are useful in other stereospecific alkene syntheses involving copper(I), particularly those based on conjugate addition to α,β-unsaturated ketones.[8,9]

Two equivalents of the 1-propenyl group are used in this procedure for each equivalent of alkyl halide. Although the fate of the second propenyl group has not been investigated in detail, it presumably appears as 1-propenylcopper(I) and its decomposition products; similar behavior is observed in the conjugate addition reaction, in which only one of the two alkyl groups of a lithium dialkylcuprate is normally utilized, the second appearing, at least in part, as an alkyl-copper(I) compound. The procedure described here is designed to minimize the production of 2,4-hexadiene from thermal decomposition of 1-propenylcopper(I). In other preparations, this precaution may be unnecessary.

The choice of solvent in these couplings is dictated in part by convenience. Ether containing several equivalents of hexamethylphosphoric triamide per equivalent of organometallic reagent and tetrahydrofuran or tetrahydrofuran–ether mixtures usually give comparable yields of products; pure ether and ether–hydrocarbon mixtures are inferior. The ether–hexamethylphosphoric triamide solvent system seems easiest to work up since partitioning of the organic phase against water can be used to remove most of the inorganic salts and the hexamethylphosphoric triamide present at the conclusion of the reaction.

The most important alternative to this coupling procedure involves

TABLE I

YIELDS OF (E)-2-UNDECENE PREPARED FROM 1 EQUIVALENT OF THE INDICATED SUBSTRATES BY REACTION EITHER WITH 1 EQUIVALENT OF LITHIUM DIPROPENYLCUPRATE OR 1 EQUIVALENT OF PROPENYLLITHIUM

	Reaction with R₂CuLi				Reaction with RLi			
Substrate	Yield (%)[a]	Time (hours)	Temperature (°)	Solvent	Yield (%)[a]	Time (hours)	Temperature (°)	Solvent
1-Iodooctane	100	0.75	−30 to −20	Ether + 4 equivalents HMPA	76	0.25	−35	Et₂O + 4 equivalents HMPA
	100	0.75	−30 to −25	THF	100	1.7	+25	THF or DME
	73	18	+25	Et₂O	7	23	+25	Et₂O
1-Bromooctane	100	3	−30 to +5	Et₂O + 4 equivalents HMPA	100	3.5	+25	THF
1-Chlorooctane[b]	80	48	+25	Et₂O + 4 equivalents HMPA	50	70	+25	DME
1-Tosyloctane	95	2.5	−10	Et₂O + THF[c]	0	48	+25	THF

[a] Yields determined by gas chromatography (Note 20). Preparative scale yields are usually approximately 10% lower.
[b] Unreacted chloride remains at the end of the reaction (with R₂CuLi: 15%; with RLi: 40%); 1.5 equivalents of R₂CuLi per equivalent of chloride gives 95% (E)-2-undecene after 5 days at 25°.
[c] A volume of THF equal to that of the ether present was added after formation of the cuprate was complete. THF: tetrahydrofuran; HMPA: hexamethylphosphoric triamide; Et₂O: diethyl ether; DME: 1,2-dimethoxyethane.

111

TABLE II[12]

REACTIONS OF PRIMARY AND SECONDARY TOSYLATES WITH ORGANOCUPRATES

$$R—OTs + R_2'CuLi \rightarrow R—R'$$

R	R'	Equivalents of Cuprate	Temper- ature (°)	Time (hours)	R—R' Yield (%)
$CH_3(CH_2)_6CH_2—$	$CH_3—$	2	0	1	95
$CH_3(CH_2)_6CH_2—$	$CH_3CH_2CHCH_3$	5	—15	16	85
$CH_3(CH_2)_5CHCH_3$	$CH_3—$	2	—20	4	87
$(CH_3)_3CCH_2—$	$C_6H_5—$	3	25	72	80
$c\text{-}C_5H_9—$	$CH_3—$	2	0	5	65
$c\text{-}C_6H_{11}—$	$CH_3—$	2	0	5	20

the reaction of organolithium reagents with organic halides (Table I).[10] A similar coupling reaction of organomagnesium reagents in tetrahydrofuran or in hexamethylphosphoric triamide has been reported.[11] Procedures using pure organolithium reagents are potentially capable of higher yields of products based on the organolithium reagent, since all of the organolithium moieties present in solution are available for reaction; further, they avoid thermally unstable organocopper(I) intermediates. However procedures based on lithium divinylcuprates are compatible with a wider range of functional groups present in the organic halide.

The general procedure described here is applicable to the preparation and use of many lithium dialkylcuprates derived from the thermally less stable *primary* alkyllithium reagents, although the solutions of intermediate organocopper compounds are usually highly colored and may contain suspended copper metal or organocopper compounds. Applications involving *secondary* organolithium reagents frequently fail as a result of thermal decomposition of the intermediate alkylcopper(I) compounds; these reactions may require the presence of stabilizing phosphine ligands in solution, although the work-up of reaction mixtures containing phosphines is normally more difficult than that described here. Several examples shown in Table II illustrate reaction of primary and secondary tosylates with various organocuprates, using diethyl ether as solvent.

1. Department of Chemistry, Massachusetts Institute of Technology, Cambridge, Massachusetts 02139. This work was supported in part by grants from the National Science Foundation and the International Copper Research Association, Inc.
2. Fellow of the Centre National de la Recherche Scientifique, Paris, France.
3. G. M. Whitesides, C. P. Casey, and J. K. Krieger, *J. Amer. Chem. Soc.*, **93**, 1379 (1971), and references cited therein.
4. (a) G. B. Kauffman and L. A. Teter, *Inorg. Syn.*, **7**, 9 (1963). (b) J. G. Smith and R. T. Wikman, *Syn. React. Inorg. Metal–Org. Chem.*, **4**, 239 (1974).
5. G. Zweifel and C. C. Whitney, *J. Amer. Chem. Soc.*, **89**, 2753 (1967); H. C. Brown, D. H. Bowman, S. Misumi, and M. K. Unni, *J. Amer. Chem. Soc.*, **89**, 4531 (1967); G. Zweifel and H. Arzoumanian, *J. Amer. Chem. Soc.*, **89**, 5086 (1967); E. J. Corey, J. I. Shulman, and H. Yamamoto, *Tetrahedron Lett.*, 447 (1970); J. F. Normant and M. Bourgain, *Tetrahedron Lett.*, 2583 (1971).
6. G. M. Whitesides, W. F. Fisher, Jr., J. San Filippo, Jr., R. W. Bashe, and H. O. House, *J. Amer. Chem. Soc.*, **91**, 4871 (1969).
7. In small-scale experiments, overnight reaction of lithium dipropenylcuprate with iodobenzene in ether containing 20 equivalents of pyridine at 25° gave 1-propenylbenzene in 60% yield. For coupling with aromatic halides, this solvent system is superior either to ether–tetrahydrofuran or to ether containing 4 equivalents of hexamethylphosphoric triamide.
8. J. Hooz and R. B. Layton, *Can. J. Chem.*, **48**, 1626 (1970); F. Näf and P. Degen, *Helv. Chim. Acta*, **54**, 1939 (1971); F. Näf, P. Degen, and G. Ohloff, *Helv. Chim. Acta*, **55**, 82 (1972); E. J. Corey and R. L. Carney, *J. Amer. Chem. Soc.*, **93**, 7318 (1971).
9. J. F. Normant, *Synthesis*, 63 (1972).
10. G. Linstruwelle, *Tetrahedron Lett.*, 3809 (1974).
11. J. F. Normant, *Bull. Soc. Chim. Fr.*, 1888 (1963); H. Normant, *Angew. Chem. Int Ed. Engl.*, **6**, 1046 (1967); *Bull. Soc. Chim. Fr.*, 791 (1968).
12. C. R. Johnson and G. A. Dutra, *J. Amer. Chem. Soc.*, **95**, 7777 (1973).

PREPARATION OF *N*-AMINOAZIRIDINES: *trans*-1-AMINO-2,3-DIPHENYLAZIRIDINE, 1-AMINO-2-PHENYLAZIRIDINE, AND 1-AMINO-2-PHENYLAZIRIDINIUM ACETATE

**(1-Aziridinamine, *trans*-($\pm$)-2,3-diphenyl-,
1-aziridinamine, ($\pm$)-2-phenyl-, and
1-aziridinamine, monoacetate, ($\pm$)-2-phenyl-)**

Submitted by Robert K. Müller, Renato Joos, Dorothee Felix, Jakob Schreiber, Claude Wintner, and A. Eschenmoser[1]
Checked by Robert E. Ireland, J. Kleckner, and David M. Walba

1. Procedure

Caution! Steps B and D should be carried out in a hood behind a safety screen.

A. *trans*-2,3-*Diphenyl*-1-*phthalimidoaziridine* [*trans*-($\pm$)-2-(2,3-diphenyl-1-aziridinyl)-1*H*-isoindole-1,3(2*H*)-dione] (Note 1). A 1-l., three-necked, round-bottomed flask equipped with an efficient mechanical stirrer is charged with a mixture of 19.5 g. (0.12 mole) of *N*-aminophthalimide [2-amino-1*H*-isoindole-1,3(2*H*)-dione] (Note 2), 108g. (0.60 mole) of (*E*)-stilbene (Note 3), and 300 ml. of dichloromethane (Note 4). To the resulting suspension is added 60 g. (*ca.* 0.12 mole) of lead tetraacetate (Note 5) over a period of 10 minutes with *vigorous* stirring at room temperature. Stirring is continued for a further 30 minutes, after which time the mixture is filtered through Celite®, and the Celite® is washed twice with 50-ml. portions of dichloromethane. The combined filtrates are transferred to a 4-l. beaker, and 1.5 l. of pentane is added with gentle stirring and cooling in an ice bath. The yellow precipitate that forms after 15 minutes is suction filtered and redissolved in 200 ml. of dichloromethane. The solution thus obtained is swirled for 5 minutes with 20 g. of silica gel, filtered through Celite®, and the filter pad is washed with two 100-ml. portions of dichloromethane. To this dichloromethane solution is once again added 1.5 l. of pentane with cooling. The precipitate that has formed after 0.5 hour's standing is dried under vacuum (10 mm.) at room temperature for 5 hours, yielding 16–21 g. (39–51%) of product, m.p. 175°, which is of sufficient purity for use in the next step (Note 6).

B. *trans*-1-*Amino*-2,3-*diphenylaziridine* (Note 7). To a magnetically stirred suspension of 20.4 g. (0.060 mole) of *trans*-2,3-diphenyl-1-phthalimidoaziridine in 150 ml. of 95% ethanol in a 500-ml., round-bottomed flask at room temperature is added 150 ml. (3 moles) (Note 8) of hydrazine hydrate (Note 9). The mixture is stirred for 40 minutes while maintaining the temperature at 43–45° (Note 10) with a thermostated oil bath. The resulting cloudy yellow solution (Note 11) is cooled and filtered through Celite®. The filtrate is poured into a 2-l. separatory funnel containing 400 ml. of ether and 200 g. of ice, and the mixture is shaken vigorously. The organic phase is separated and washed with three 200-ml. portions of ice-cold water, and the aqueous washings are reextracted with a 250-ml. portion of ether (Note 12). The combined ethereal extracts are dried over anhydrous potassium carbonate, filtered through Celite® if necessary, and concentrated to approximately 300 ml. on a rotary evaporator at room temperature. Addition of 400 ml. of pentane and overnight storage at −20° leads to crystallization of 7.9–9.5 g. (63–75%) of *trans*-1-amino-2,3-diphenylaziridine (Note 13) as colorless crystals, m.p. 93–94° (decomp.) (Note 14). A further 1.2–2.2 g. (10–17%) is obtained by concentration of the mother liquor to 50–80 ml. and addition of approximately 50 ml. of pentane (Note 7).

C. *Styrene Glycol Dimesylate* (1-phenyl-1,2-ethanediol dimethylsulfonate) (Note 15). A 300-ml., three-necked, round-bottomed flask equipped with a thermometer, an efficient stirrer (Note 16), and a dropping funnel is charged with a solution of 34.5 g. (0.25 mole) of styrene glycol (Note 17) in 90 ml. of pyridine (Note 18). The solution is cooled to −5° by means of an ice-salt bath, and 44 ml. (64.7 g., 0.56 mole) of methanesulfonyl chloride (Note 19) is added dropwise over a 1-hour period, while maintaining the temperature at or below 0° (Note 20). Stirring is continued for 4 hours at 2–4°, the flask now being cooled by an ice-water bath. The reaction mixture is mixed thoroughly with 600 g. of ice, and the dimesylate precipitates. After careful acidification of the mixture with aqueous 6N hydrochloric acid to approximately pH 3 (Note 21), the dimesylate is suction filtered, washed twice with 100-ml. portions of ice water, and pressed as dry as possible. This product, which is still moist, is transferred to a separatory funnel and is shaken well with 200 ml. of dichloromethane. The dichloromethane is separated, and the aqueous layer is extracted further with two 20-ml. portions of dichloromethane. The combined dichloromethane layers are dried over anhydrous magnesium sulfate, and then 250–300 ml. of pentane is added to the solution, until crystallization just begins. After 2 hours in a deep freeze at −25°, the crystals are collected, washed with two 30-ml. portions of pentane precooled to 0°, and dried to constant weight in a vacuum desiccator (10 mm.) at room temperature, yielding 62–64 g. (84–86%) of white, crystalline dimesylate, m.p. 93–94° (Note 15).

D. *Hydrazinolysis of Styrene Glycol Dimesylate.* A 1-l. round-bottomed flask equipped with a magnetic stirring bar is charged with 50 ml. (1 mole) of hydrazine hydrate (Note 9). Finely powdered styrene glycol dimesylate (20 g., 0.068 mole), m.p. 93–94°, is then added with gentle stirring at room temperature. To the resulting slurry is slowly added 600 ml. of pentane. The stirring speed should be adjusted in such a manner that the two phases mix somewhat, but the dimesylate–hydrazine hydrate layer is not deposited on the upper walls of the flask. After 20–24 hours of stirring at room temperature two entirely clear layers can be observed upon the cessation of stirring (Note 22), indicating that the reaction is complete. The hydrazine hydrate is separated from the pentane and extracted with two 30-ml. portions of pentane. The combined pentane layers are filtered through cotton, which holds back any remaining droplets of hydrazine hydrate, into a 1-l. round-bottomed flask. At this point *Step E* is followed for 1-amino-2-phenylaziridine, and *Step F* for 1-amino-2-phenylaziridinium acetate.

E. *1-Amino-2-phenylaziridine* (Note 23). If the pentane solution from *Step D* is removed on a rotary evaporator at room temperature, 7.5–7.7 g. (82–85%) of 1-amino-2-phenylaziridine, suitable for preparative use, is obtained. Kugelrohr distillation of this material on a 1–2 g. scale (0.01 mm./60–65° oven temperature) (Note 24) gives a recovery of over 90% (Notes 23 and 25).

F. *1-Amino-2-phenylaziridinium Acetate* (Note 23). The pentane solution from *Step D* is stirred with a magnetic stirrer and cooled to 0°, and 3.9 ml. (0.068 mole) of acetic acid is measured for addition. Three drops of acetic acid are added at first, and stirring is continued at 0° until the precipitation of white 1-amino-2-phenylaziridinium acetate begins. If necessary, crystallization is initiated by scratching with a glass rod or by addition of a seed crystal from a previous run. The remainder of the acetic acid is then added over a 10-minute period, and stirring is continued for 20 minutes further, while maintaining the temperature at 0°. The salt is filtered, washed with 30 ml. of pentane precooled to 0°, and dried in a vacuum desiccator (10 mm.) at room temperature to yield 10.0–10.5 g. (76–79%) (Note 26) of product, m.p. 69–70° (Note 27) which is suitable for preparative purposes. Recrystallization is possible; it must, however, be carefully carried out to avoid the formation of a yellow product, whose melting point is lower than that of the crude product. A solution of 10 g. of 1-amino-2-phenylaziridinium acetate in 40 ml. of dichloromethane is prepared at a maximum temperature of 20–22°. The turbid solution is immediately filtered through Celite® and the filter washed with two 10-ml. portions of dichloromethane. The resulting clear solution is treated with 200–250 ml. of pentane until crystallization just commences, and then placed in a deep freeze at −25° for 2 hours. Filtration, washing with

30 ml. of pentane precooled to 0°, and drying as before, afford 9.2–9.3 g. of 1-amino-2-phenylaziridinium acetate, m.p. 70–72° (Notes 23 and 28).

2. Notes

1. *trans*-2,3-Diphenyl-1-phthalimidoaziridine is available from Fluka AG, CH-9470 Buchs.

2. *N*-Aminophthalimide is available from Fluka AG or may be prepared from phthalimide and hydrazine.[2] The quality is important; the m.p. should be 199–202° with subsequent resolidification of the melt due to thermal reaction. Recrystallization, if necessary, can be carried out in ethanol. The checkers observed that with one batch of recrystallized material, the solid never really did melt, but seemed to sinter at ∼200°.

3. Technical-grade (*E*)-stilbene obtained from Fluka AG gives satisfactory results, although a better grade is preferable. The checkers used reagent-grade material obtained from Aldrich Chemical Company, Inc.

4. This was distilled over phosphorus pentoxide.

5. "Purum"-grade lead tetraacetate, 85–90%, moistened with acetic acid, obtained from Fluka AG was used. The checkers used reagent-grade material, moistened with acetic acid, which can be purchased from Matheson Coleman and Bell.

6. The yield of the reaction, while always at least 39%, is subject to fluctuation. The product may be contaminated with small amounts of (*E*)-stilbene and/or lead salts. The presence of (*E*)-stilbene can easily be monitored by thin layer chromatography, using ready-prepared Silica Gel F_{254} plates available from E. Merck & Company, Darmstadt, Germany. The plates are developed with dichloromethane, and the spots detected under ultraviolet light. In runs of smaller scale, or if product of higher purity is desired, column chromatography on silica gel may replace the work-up described; the reaction mixture is filtered through Celite®, and the resulting solution is concentrated on a rotary evaporator. The residue is then chromatographed on 570 g. of silica gel. Eluting with dichloromethane gives the stilbene first, then small amounts of unidentified impurities, and finally the desired adduct in 70–73% yield.

7. *trans*-1-Amino-2,3-diphenylaziridine decomposes thermally, largely to (*E*)-stilbene and nitrogen. In the crystalline state it is stable

for several hours at room temperature, and for several weeks (probably for several months) at −20°. It decomposes within 3 days at room temperature in aprotic solvents and much more rapidly in protic solvents.[3] Decomposition is still faster in the presence of traces of acids. The work-up described should be completed as quickly and as *precisely* as possible, and the product should be stored in a deep freeze if it is not to be used immediately. For the use of this reagent in the α,β-epoxyketone fragmentation, see references 3 and 4.

8. Use of lesser amounts of hydrazine hydrate or ethanol causes precipitation of the product as a pasty mass that dissolves only slowly in ether, thereby making the work-up difficult.

9. "Purum"-grade hydrazine hydrate obtained from Fluka AG was used. The checkers used 99–100% hydrazine hydrate purchased from Matheson Coleman and Bell.

10. The reaction temperature must be carefully controlled. At temperatures above 48° the yield is markedly reduced by decomposition of the product, and below 43° the reaction time is greatly lengthened.

11. The submitters report that if very pure *trans*-2,3-diphenyl-1-phthalimidoaziridine is used, no insoluble matter is present at this point, and filtration through Celite® is not necessary.

12. Before working up the reaction mixture, it is recommended to test whether the reaction is complete by thin layer chromatography (Note 6). If (*E*)-stilbene is observed, the reaction should be interrupted immediately. The checkers found that the work-up was complicated by formation of emulsions. Small quantities of brine were used to aid separation of the phases.

13. The product has the following spectral properties; infrared (chloroform) cm.$^{-1}$: 3340, 1603, 1495, 1450, 1085, 1070, 1030 (a weak band at 960 cm.$^{-1}$ is due to (*E*)-stilbene, as there is always some decomposition of the *trans*-1-amino-2,3-diphenylaziridine during the recording of the spectrum) (Note 7); proton magnetic resonance (chloroform-*d*) δ, multiplicity, number of protons, assignment, coupling constant *J* in Hz.: 3.10 (broad multiplet, 2, NH_2), 3.22 and 3.36 (AB quartet, 2, *J* = 5), 7.1–7.6 (multiplet, 10). The nonequivalence of the two methine protons is due to slow inversion at nitrogen, and confirms the *trans*-2,3-substitution.

14. The melting point tube is placed in the bath at 85° and heated rapidly.

15. A second crystallization from dichloromethane–pentane is sometimes necessary to achieve material having this melting point. Styrene glycol dimesylate must be stored in a refrigerator, since slow decomposition takes place at room temperature.

16. Toward the end of the reaction, the mixture becomes quite viscous. Unless the stirring assembly is capable of mixing material at the flask walls, homogeneous temperature control cannot be guaranteed.

17. Styrene glycol is available from Aldrich Chemical Company, Inc. or from Eastman Organic Chemicals. Alternatively, it may be prepared by hydrolysis of styrene oxide.[3] If the glycol melts at lower than 63°, it should be recrystallized before use.

18. Dried over potassium hydroxide and distilled.

19. "Purum"-grade methanesulfonyl chloride supplied by Fluka AG or 98%-pure material supplied by Eastman Organic Chemicals was used.

20. The addition rate is about 30 drops/minute. After about half of the methanesulfonyl chloride has been added, white crystals of pyridine hydrochloride begin to precipitate, and the solution becomes viscous (Note 16).

21. About 110–120 ml. of aqueous $6N$ hydrochloric acid is needed. The temperature should not be allowed to rise above 5°.

22. The rate of the reaction is influenced by the speed of stirring. At slow speeds, 30 hours may be required for completion, without, however, any lowering of the final yield. The role of pentane is to continuously remove newly formed product from the hydrazine solvent.

23. 1-Amino-2-phenylaziridine decomposes at temperatures over 0°, and must therefore be stored in a deep freeze at −25°, at which temperature it is crystalline. 1-Amino-2-phenylaziridinium acetate is somewhat more stable, but it too decomposes within 2 days at room temperature. It can be kept unchanged in a deep freeze for months. For the use of these two reagents in the α,β-epoxyketone fragmentation, see references 3 and 4.

24. In order to minimize decomposition, the distillation should be carried out on small portions at the lowest possible temperature and pressure.

25. The distilled product has the following proton magnetic resonance spectrum (chloroform-d) δ, number of protons, assignment:

1.91, 1.98, and 2.06 (2, AB part of ABX), 2.50, 2.58, 2.64, and 2.72 (1, X part of ABX), 3.60 (2, NH_2), 7.25 (5). Undistilled material shows substantially the same spectrum.

26. By the addition of 2 further drops of acetic acid to the mother liquor, and overnight cooling at $-25°$, an additional 0.3–0.4 g. (2–3%) of product, m.p. 60–62°, can be isolated.

27. For the melting point determination the capillary is placed in the apparatus at 60°, and the temperature is raised at 4°/minute.

28. Recrystallized 1-amino-2-phenylaziridinium acetate has the following proton magnetic resonance spectrum (chloroform-*d*) δ, number of protons, assignment: 1.95–2.20 (5, AB part of ABX and CH_3 at δ 2.02), 2.67–2.95 (1, X part of ABX), 6.50–6.70 (3, NH_3) 7.0–7.4 (5).

3. Discussion

trans-1-Amino-2,3-diphenylaziridine and 1-amino-2-phenylaziridine are reagents for the α,β-epoxyketone $\rightarrow$ alkynone fragmentation,[3] an example of which is given in this volume.[4] An alternative preparation of 1-amino-2-phenylaziridine is by the hydrazinolysis of 2-phenyl-1-phthalimidoaziridine.[3]

The lead tetraacetate reaction between *N*-aminophthalimide and (*E*)-stilbene was first described by Rees,[5] and the hydrazinolysis of the addition product by Carpino.[6] The procedures described here incorporate their methods, with improvements. The dimesylate–hydrazine reaction was first described by Paulsen[7] in the carbohydrate series.

1. Laboratorium für Organische Chemie, Eidgenössische Technische Hochschule, CH-8006 Zürich, Switzerland.
2. H. D. K. Drew and H. H. Hatt, *J. Chem. Soc. London*, 16 (1937).
3. D. Felix, R. K. Müller, U. Horn, R. Joos, J. Schreiber, and A. Eschenmoser, *Helv. Chim. Acta*, **55**, 1276 (1972).
4. D. Felix, C. Wintner, and A. Eschenmoser, *Org. Syn.*, **55**, 52 (1975).
5. D. J. Anderson, T. L. Gilchrist, D. C. Horwell, and C. W. Rees, *J. Chem. Soc. C*, 576 (1970).
6. L. A. Carpino and R. K. Kirkley, *J. Amer. Chem. Soc.*, **92**, 1784 (1970).
7. H. Paulsen and D. Stoye, *Angew. Chem.*, **80**, 120 (1968); *Angew. Chem. Int. Ed. Engl.*, **7**, 134 (1968); *Chem. Ber.*, **102**, 820 (1969).

SECONDARY AND TERTIARY ALKYL KETONES FROM CARBOXYLIC ACID CHLORIDES AND LITHIUM PHENYLTHIO(ALKYL)CUPRATE REAGENTS: *tert*-BUTYL PHENYL KETONE

(1-Propanone, 2,2-dimethyl-1-phenyl)

$$C_6H_5SLi + CuI \xrightarrow[\text{tetrahydrofuran}]{25°} C_6H_5SCu + LiI$$

$$C_6H_5SCu + (CH_3)_3CLi \xrightarrow[\substack{\text{tetrahydrofuran,}\\\text{pentane}}]{-60° \text{ to } -65°} C_6H_5S[(CH_3)_3C]CuLi$$

$$C_6H_5S[(CH_3)_3C]CuLi + C_6H_5COCl \xrightarrow[\text{tetrahydrofuran}]{-60° \text{ to } -65°} (CH_3)_3CCOC_6H_5$$

Submitted by Gary H. Posner and Charles E. Whitten[1]
Checked by Joyce M. Wilkins and Herbert O. House

1. Procedure

Caution! Since the odor of the thiophenol (benzenethiol) used in this preparation is unpleasant, both steps of this preparation should be conducted in a hood and the glassware used should be washed before it is removed from the hood.

A. *Lithium Phenylthio(tert-butyl)cuprate.* A dry, 200-ml., round-bottomed flask is fitted with a magnetic stirring bar and a 100-ml. pressure-equalizing dropping funnel, the top of which is connected to a nitrogen inlet. After the apparatus has been flushed with nitrogen, 50 ml. of $1.60M$ (0.080 mole) butyllithium (Note 1) solution is placed in the flask and cooled with an ice bath. Under a nitrogen atmosphere, a solution of 8.81 g. (0.080 mole) of freshly distilled thiophenol (Note 2) in 30 ml. of anhydrous tetrahydrofuran (Note 3) is added dropwise to the cooled, stirred solution. An aliquot of the resulting solution (Note 4) is standardized by quenching in water followed by titration with aqueous $0.10N$ hydrochloric acid to a green end point with a bromocresol indicator. The concentration of lithium thiophenoxide prepared in this manner is typically $1.0M$.

A dry, 250-ml., three-necked, round-bottomed flask is equipped with a sealed mechanical stirrer (Note 5), a glass stopper, and a rubber septum through which are inserted hypodermic needles with which to evacuate the flask and to admit nitrogen. After the apparatus has been

122

flushed with nitrogen, 4.19 g. (0.022 mole) of purified copper(I) iodide (Note 6) is added, and while warming with a flame, the apparatus is evacuated and then refilled with nitrogen. After this procedure has been performed twice, the flask is allowed to cool, the stopper is replaced with a thermometer, and 45 ml. of anhydrous tetrahydrofuran is added (Note 3) with a hypodermic syringe. With continuous stirring, 22 ml. of 1.0M (0.022 mole) lithium thiophenoxide solution is added from a syringe to the slurry of copper(I) iodide. After 5 minutes, the resulting yellow solution is cooled, with continuous stirring, to $-65°$ by use of an acetone–dry ice cooling bath. Some copper(I) thiophenoxide usually separates from solution at approximately $-45°$. When the temperature of the mixture has reached approximately $-65°$, 13.6 ml. (0.0218 mole) of 1.60M *tert*-butyllithium (Note 7) solution is added from a syringe to the stirred mixture at such a rate that the temperature of the mixture remains at $-60°$ to $-65°$. The resulting cloudy yellow–orange solution of the cuprate reagent is stirred at $-60°$ to $-65°$ for 5 minutes (Note 8).

B. *tert-Butyl Phenyl Ketone.* By means of a syringe a solution of 2.81 g. (0.020 mole) of freshly distilled benzoyl chloride (Note 9) in 15 ml. of anhydrous tetrahydrofuran (Note 3) is added dropwise with stirring to the solution of the cuprate reagent at $-60°$ to $-65°$. The resulting yellow–brown solution is stirred for 20 minutes at $-60°$ to $-65°$ and then quenched by the addition, from a syringe, of 5 ml. of anhydrous methanol. The red–orange reaction mixture is allowed to warm to room temperature and is then poured into 100 ml. of aqueous saturated ammonium chloride. The copious precipitate of copper(I) thiophenoxide is separated by suction filtration, and the precipitate is washed thoroughly with several 50-ml. portions of ether. The combined filtrate is further extracted with three 100-ml. portions of ether. The combined ethereal solution is washed with two 50-ml. portions of aqueous 1N sodium hydroxide and with one 50-ml. portion of aqueous 2% sodium thiosulfate. Each of the aqueous washes is extracted in turn with a fresh 50-ml. portion of ether. The combined ethereal solution is dried with anhydrous magnesium sulfate, filtered, and then concentrated by distillation through a short Vigreux column. The residual pale yellow liquid (Note 10) is distilled through a short column under reduced pressure to yield 2.73–2.82 g. (84–87%) of *tert*-butyl phenyl ketone as a colorless liquid b.p. 105–106° (15 mm.), 114–115° (44 mm.), n^{20} D 1.5092, n^{25} D 1.5066 (Note 11).

2. Notes

1. Solutions containing approximately $1.6M$ butyllithium in hexane were purchased either from Alfa Inorganics, Inc., or from Foote Mineral Company. The concentration of butyllithium in these solutions can be determined either by a double titration procedure[2] or by dilution with anhydrous tetrahydrofuran followed by titration with 2-butanol in the presence of a 2,2'-bipyridyl indicator.[3] In either case the total base concentration in the reagent is determined by titration with standard aqueous acid.

2. Thiophenol, purchased from Aldrich Chemical Company, Inc., was redistilled before use; b.p. 65–66° (42 mm.).

3. Commercial anhydrous tetrahydrofuran was distilled from lithium aluminum hydride and stored under nitrogen.

4. The submitters report that this solution may be stored under nitrogen at 0° for several days without deterioration.

5. Although the submitters had recommended use of a magnetic stirring bar, the checkers encountered considerable difficulty in maintaining adequate stirring of the cold reaction mixture with a magnetic stirrer and recommend use of a sealed mechanical stirrer such as a Truebore® stirrer.

6. Copper(I) iodide, purchased from Fisher Scientific Company, was purified by continuous extraction with anhydrous tetrahydrofuran in a Soxhlet extractor for approximately 12 hours in order to remove colored impurities. The residual copper(I) iodide was then dried under reduced pressure at 25° and stored under nitrogen in a desiccator.

7. A pentane solution of tert-butyllithium (purchased from either Alfa Inorganics, Inc. or Lithium Corporation of America, Inc.) was standardized by one of the previously described titration procedures (Note 1). If possible, it is desirable to use a freshly opened bottle of tert-butyllithium since previously used bottles of this reagent often contain lithium tert-butoxide which will lead to formation of a contaminant in the final product (Note 10).

8. Although the submitters report that this reagent is stable at 0° (i.e., still reactive toward benzoyl chloride) for periods of at least one hour under a nitrogen atmosphere,[4] the checkers repeatedly observed evidence of thermal decomposition when the solution was allowed to warm above −40°. This decomposition was indicated by the appearance of a red–brown coloration as the reagent was warmed to −40°; as the

temperature was raised further to $-25°$ and to $0°$, the mixture progressively exhibited a darker brown color.

9. Benzoyl chloride (purchased from Eastman Organic Chemicals) was redistilled before use; b.p. 35–36° (0.5 mm.).

10. The checkers found that with previously opened bottles of *tert*-butyllithium, the crude product was often contaminated with *tert*-butyl benzoate (from lithium *tert*-butoxide; see Note 7). The presence of this impurity in the crude product may be detected either by the presence of an extra infrared peak at 1720 cm.$^{-1}$ (conjugated ester), or by gas chromatographic analysis. On a 1.3-m. gas chromatographic column, packed with silicone fluid, No. SE-52, suspended on Chromosorb P and operated at 155°, the retention time of *tert*-butyl phenyl ketone was 4.4 minutes, and the retention times of potential impurities, methyl benzoate and *tert*-butyl benzoate were 2.4 minutes and 7.8 minutes, respectively. If a small amount of *tert*-butyl benzoate is present in the crude product, it is most easily removed by heating a mixture of the crude product with 1% by weight of *p*-toluenesulfonic acid (4-methylbenzenesulfonic acid) on a steam bath for 10 minutes followed by partitioning the product between ether and aqueous sodium bicarbonate. After the resulting ether solution has been dried and distilled, pure *tert*-butyl phenyl ketone is obtained.

11. The product exhibits a single gas chromatographic peak (see Note 10). The spectral properties of the product are as follows; infrared (carbon tetrachloride) cm.$^{-1}$: 1680 (conjugated ketone), 1395 and 1370 [$C(CH_3)_3$]; proton magnetic resonance (carbon tetrachloride) δ, multiplicity, number of protons, assignment: 1.30 [singlet, 9, $C(CH_3)_3$], 7.2–7.9 (multiplet, 5, C_6H_5); ultraviolet (95% ethanol) nm. max. (ϵ): 237 (7350) and 272 (620); mass spectrum m/e (relative intensity): 162 (M, 45), 106 (28), 105 (100), 77 (63), 57 (40), 51 (23), and 41 (30). The physical constants reported for the product are: b.p. 103–104° (13 mm.),[5] n^{20} D 1.5090.[6]

3. Discussion

tert-Butyl phenyl ketone has been prepared by the reactions of benzoic acid with *tert*-butyllithium,[7,8] of acetophenone (1-phenylethanone) with iodomethane and base,[5,9] of benzaldehyde with *tert*-butylmagnesium chloride followed by oxidation,[10] and of 2,2-dimethylpropanoyl chloride with phenylmagnesium bromide.[11]

The procedure described here illustrates the preparation of mixed lithium arylhetero(alkyl)cuprate reagents and their reactions with carboxylic acid chlorides.[4] These mixed cuprate reagents also react with α,α'-dibromoketones,[12] primary alkyl halides,[4] and α,β-unsaturated ketones,[4] with selective transfer of only the alkyl group.

Two limitations on the broad utility of organocopper reagents have often been the difficulty in using thermally unstable lithium sec- and especially tert-alkylcuprates[13] and the need for a large (e.g., 300–500%) excess of an organocuprate to achieve complete conversion of a substrate to a product. Both of these limitations are circumvented by using lithium phenylthio(tert-alkyl)cuprates, which react with approximately equimolar amounts of carboxylic acid chlorides to form the corresponding tert-alkyl ketones in high yield, even with the yield based on the transferred alkyl group. Furthermore, this alkyl group transfer can be achieved in the presence of other functional groups (e.g., remote halogen or ester functionalities) in the carboxylic acid chloride substrate (Equation 1). Transfer of secondary alkyl groups can also be accomplished efficiently in this way (Equation 2).

$$C_2H_5O_2CCH_2CH_2COCl \xrightarrow[\text{tetrahydrofuran, } -78°, \text{ 15 minutes}]{1.2 \text{ equivalent } C_6H_5S(tert\text{-}C_4H_9)CuLi}$$

$$C_2H_5O_2CCH_2CH_2COC_4H_9\text{-}tert \quad (1)$$
$$(65\%)$$

$$C_6H_5COCl \xrightarrow[\text{tetrahydrofuran, } -78°, \text{ 15 minutes}]{1.3 \text{ equivalent } C_6H_5S(sec\text{-}C_4H_9)CuLi} C_6H_5COC_4H_9\text{-}sec \quad (2)$$
$$(80\%)$$

The reaction of tert-alkyl Grignard reagents with carboxylic acid chlorides in the presence of a copper catalyst provides tert-alkyl ketones in substantially lower yields than those reported here.[4,14] The simplicity and mildness of experimental conditions and isolation procedure, the diversity of substrate structural type, and the functional group selectivity of these mixed organocuprate reagents render them very useful for conversion of carboxylic acid chlorides to the corresponding secondary and tertiary alkyl ketones.[15]

1. Department of Chemistry, the Johns Hopkins University, Baltimore, Maryland 21218; this work was supported by the National Science Foundation (GP-33667).
2. G. M. Whitesides, C. P. Casey, and J. K. Krieger, J. Amer. Chem. Soc., 93, 1379 (1971).
3. M. Gall and H. O. House, Org. Syn., 52, 39 (1972).
4. G. H. Posner, C. E. Whitten, and J. J. Sterling, J. Amer. Chem. Soc., 95, 7788 (1973).
5. A. Haller and E. Bauer, C.R.H. Acad. Sci., 148, 73 (1909).

6. C. Cherrier and J. Metzger, *C.R.H. Acad. Sci.*, **226**, 797 (1948).
7. Unpublished results of C. H. Heathcock and R. Radcliff as reported in Ref. 8.
8. For a general discussion of ketone formation from carboxylic acids and organolithium reagents, see M. J. Jorgenson, *Org. React.*, **18**, 1 (1970).
9. J. U. Nef, *Justus Liebigs Ann. Chem.*, **310**, 316 (1900).
10. A. Favorskii, *Bull. Soc. Chim. Fr.*, **3**, 239 (1936).
11. J. Thiec, *Ann. Chim. Paris*, **9**, 51 (1954).
12. G. H. Posner and J. J. Sterling, *J. Amer. Chem. Soc.*, **95**, 3076 (1973).
13. G. M. Whitesides, W. F. Fischer, Jr., J. San Filippo, Jr., R. W. Bashe, and H. O. House, *J. Amer. Chem. Soc.*, **91**, 4871 (1969).
14. J. E. Dubois, M. Boussu, and C. Lion, *Tetrahedron Lett.*, 829 (1971) and references cited therein; J. A. MacPhee and J. E. Dubois, *Tetrahedron Lett.*, 467 (1972).
15. For use of other organocopper reagents in converting carboxylic acid chlorides to ketones, see G. H. Posner and C. E. Whitten, *Tetrahedron Lett.*, 1815 (1973); G. H. Posner, C. E. Whitten, and P. E. McFarland, *J. Amer. Chem. Soc.*, **94**, 5106 (1972). For a recent report on direct and convenient preparation of lithium phenylthio(alkyl)-cuprate reagents, see G. H. Posner, D. J. Brunelle, and L. Sinoway, *Synthesis*, 662 (1974).

SULFIDE CONTRACTION *via* ALKYLATIVE COUPLING: 3-METHYL-2,4-HEPTANEDIONE

$$2(CH_3)_2NCH_2CH_2CH_2MgCl + C_6H_5PCl_2 \xrightarrow{\text{tetrahydrofuran}} C_6H_5P[CH_2CH_2CH_2N(CH_3)_2]_2$$

$$CH_3CH_2CH_2COSH + CH_3\overset{\overset{\text{O}}{\|}}{C}\underset{\underset{\text{Br}}{|}}{CH}CH_3 \xrightarrow{(C_2H_5)_3N} CH_3CH_2CH_2CO\text{—}S\text{—}\underset{\underset{\text{CH}_3}{|}}{CH}COCH_3$$

$$CH_3CH_2CH_2CO\text{—}S\text{—}\underset{\underset{\text{CH}_3}{|}}{CH}COCH_3 + C_6H_5P[CH_2CH_2CH_2N(CH_3)_2]_2 \xrightarrow[\substack{\text{acetonitrile,} \\ 70°}]{\text{LiBr}}$$

$$CH_3CH_2CH_2CO\underset{\underset{\text{CH}_3}{|}}{CH}COCH_3 + C_6H_5\overset{\overset{\text{S}}{\|}}{P}[CH_2CH_2CH_2N(CH_3)_2]_2$$

Submitted by P. LOELIGER and E. FLÜCKIGER[1]
Checked by K. MATSUO and G. BÜCHI

1. Procedure

Caution! To avoid exposure to toxic phenylphosphonous dichloride vapors, the Grignard reaction should be conducted in a hood.

A. *Bis(3-dimethylaminopropyl)phenylphosphine* [3,3′-(phenylphos-phinidene)bis(N,N-dimethyl)-1-propanamine]. A 3-l. separatory funnel is charged with 395 g. (2.5 moles) of 3-chloro-N,N-dimethyl-1-propanamine hydrochloride (Note 1), and a cold solution of 179 g. of potassium hydroxide in 540 ml. of water is added. The mixture is extracted three times with 300-ml portions of ether–dichloromethane (5:1). The organic extracts are washed with 300 ml. of aqueous 2N potassium hydroxide, combined, and dried over anhydrous sodium sulfate. The solvent is removed by distillation through a 25-cm. Vigreux column at atmospheric pressure, and the residual liquid is distilled under reduced pressure through a 13-cm. Vigreux column to give 263–276 g. (87–91%) of 3-chloro-N,N-dimethyl-1-propanamine as a colorless liquid, b.p. 72–73° (100 mm.) (Notes 2 and 3), which is used immediately in the Grignard reaction (Note 4).

A 3-l., four-necked, round-bottomed flask equipped with a sealed mechanical stirrer, a pressure-equalizing dropping funnel, a thermometer, and a condenser fitted with a nitrogen-inlet tube is charged with 48.6 g. (2.0 g.-atoms) of magnesium turnings (Note 5). The flask is flushed with dry nitrogen and thoroughly dried by means of a heat gun, and 300 ml. of anhydrous tetrahydrofuran (Note 6) is added. The Grignard reaction is initiated by adding about 10% of a solution of 243.0 g. (2.0 moles) of 3-chloro-N,N-dimethyl-1-propanamine in 300 ml. of anhydrous tetrahydrofuran (Note 6), and 4 ml. of bromoethane while gently heating the flask with the drier (Note 7). The remainder of the 3-chloro-N,N-dimethyl-1-propanamine solution is added over a period of approximately 1 hour so as to maintain gentle reflux. The reaction mixture is heated at reflux for 3 hours, after which time most of the magnesium has reacted. The dark grey solution is cooled to 0°, and a solution of 82 ml. (107.3 g., 0.60 mole) of phenylphosphonous dichloride (Note 8) in 200 ml. of anhydrous tetrahydrofuran (Note 6) is added dropwise, with efficient stirring, over a 1 hour period so that the temperature does not exceed 5° (Note 9). A greenish precipitate is formed locally where the phosphine is added. After the addition is complete, the reaction mixture is stirred and heated at reflux for 2 hours, during which time a heavy greenish precipitate is formed. After cooling to room temperature, 600 ml. of ether (Note 10) is added, and the reaction mixture is left to stand overnight, during which time the precipitate separates to the bottom of the flask. The solution is decanted into a 3-l. separatory funnel containing 300 ml. of aqueous

40% potassium hydroxide and 1 kg. of ice. The remainder of the reaction product is suction filtered with the aid of 1200 ml. of ether–dichloromethane (5:1) through a 3-cm. layer of Celite® (Note 11). The filtrate is added to the separatory funnel, and the organic layer is separated and washed twice with 600-ml. portions of aqueous saturated sodium chloride. The aqueous layer is extracted four times with 700-ml. portions of ether–dichloromethane (5:1). The combined organic extracts are dried over anhydrous sodium sulfate, and the solvent is removed on a rotary evaporator under reduced pressure. The crude yellow oil is distilled at high vacuum through a 14-cm. Vigreux column to yield 109–116 g. (65–69%, based on phenylphosphonous dichloride) of bis(3-dimethylaminopropyl)phenylphosphine as a colorless liquid, b.p. 100–108° (0.005 mm.) (Note 12). Redistillation furnishes 94–97 g. (56–58%) of product, b.p. 102–105° (0.005 mm.) (Note 13), n^{24} D 1.5265.

B. *S-(2-Oxobut-3-yl) Butanethioate.* A 750-ml., four-necked, round-bottomed flask equipped with a sealed mechanical stirrer, a pressure-equalizing dropping funnel, a thermometer, and a condenser fitted with a nitrogen-inlet tube is charged with 10.4 g. (0.10 mole) of thiobutyric acid (butanethioic acid) (Note 14) in 300 ml. of anhydrous ether (Note 10). With stirring, 10.1 g. (0.10 mole) of triethylamine (Note 15) is added all at once. Over a 15-minute period (Note 16), 15.1 g. (0.10 mole) of 3-bromo-2-butanone (Note 17) is added dropwise from the dropping funnel. The solution is heated at reflux with stirring for 1.5 hours, filtered through Celite®, and the precipitate is washed with 60 ml. of ether. The ether solution is concentrated on a rotary evaporator. The residual orange–yellow oil is dissolved in 20 ml. of benzene–ether (5:1) and filtered through 70 g. of silica gel (Note 18) using 500 ml. of this solvent mixture as eluent. The solvent is removed on a rotary evaporator under reduced pressure to furnish 17.0–17.4 g. (98–100%) of the thiol ester as a pale yellow oil which was used without further purification in the next step. (Notes 19 and 20).

C. *3-Methyl-2,4-heptanedione.* A dry, 500-ml., three-necked, round-bottomed flask equipped with a magnetic stirring bar, a pressure-equalizing dropping funnel, a thermometer, and a condenser fitted with a nitrogen-inlet tube is charged with 17.8 g. (0.20 mole) of anhydrous lithium bromide (Note 21). Under a nitrogen atmosphere, 34.8 g. (0.20 mole) of *S*-(2-oxobut-3-yl) butanethioate dissolved in 120 ml. of anhydrous acetonitrile (Note 22) is added to the flask.

With stirring, the mixture is heated with a drier until a homogeneous solution is obtained. From the dropping funnel, 69 ml. (67 g., 0.24 mole) of redistilled bis(3-dimethylaminopropyl)phenylphosphine is added all at once to the warm (*ca.* 60°) solution. The temperature rises to about 70°, and after 1–2 minutes a thick white precipitate appears. The reaction mixture is stirred at 70° for 15 hours (Note 23). After cooling to room temperature, the reaction mixture is transferred with 600 ml. of ether–dichloromethane (5:1) into a separatory funnel containing 900 ml. of cold aqueous 1N hydrochloric acid. The organic layer is separated and washed three times with 500-ml. portions of aqueous saturated sodium chloride. The aqueous phase is washed twice with 600-ml. portions of ether–dichloromethane (5:1). The combined organic layer is dried over anhydrous sodium sulfate, and the solvent removed on a rotary evaporator under reduced pressure, the temperature of the bath not exceeding 30°. The crude yellow oil is distilled through a 10-cm. Vigreux column under reduced pressure to yield 23.5–24.7 g. (83–87%) of 3-methyl-2,4-heptanedione as a colorless liquid b.p. 74–76° (9 mm.), n^{23} D 1.4455 (Note 24).

2. Notes

1. 3-Chloro-N,N-dimethyl-1-propanamine hydrochloride was purchased from Fluka AG CH–9470 Buchs or Aldrich Chemical Company, Inc.

2. The fractions may be analyzed by gas chromatography for absence of solvent. For gas chromatographic analysis a 300 cm. by 0.3 cm. glass column packed with XE-60 (1.5% w/w) coated on Chromosorb G AW DCMS (80/100 mesh) was employed.

3. The spectral properties of the product are as follows; infrared (neat) cm.$^{-1}$: 1470, 1465, 1265, 1040; proton magnetic resonance (chloroform-d) δ, multiplicity, number of protons, coupling constant J in Hz.: 3.6 (triplet, 2, $J = 6$), 1.8–2.6 (multiplet 4), 2.2 (singlet, 6).

4. It is advisable to distil the solvent one day, store the residue overnight under nitrogen at 0°, and distil the product the next morning to allow ample time for the following Grignard reaction. On standing at room temperature a white solid precipitates.

5. Magnesium turnings were purchased from E. Merck & Company, Inc., Darmstadt, Germany or J. T. Baker Chemical Company.

6. Tetrahydrofuran (purchased from Fluka AG or J. T. Baker Chemical Company) is distilled from sodium hydride prior to use. For warnings regarding the purification of tetrahydrofuran, see *Org. Syn.*, Coll. Vol. **5**, 976 (1973).

7. It is advisable to have available an ice bath for cooling, should the reaction become violent.

8. Phenylphosphonous dichloride was purchased from Fluka AG or Aldrich Chemical Company, Inc.

9. The reaction is very exothermic and cooling with an ice–sodium chloride bath is necessary.

10. Anhydrous ether was purchased from Fluka AG or J. T. Baker Chemical Company.

11. During this operation, the reaction vessel is washed with ether–dichloromethane (5:1) several times.

12. The colorless forerun weighs 12–20 g.; the dark brown residue weighs 13–22 g.

13. The reported b.p. is 102–105° (0.005 mm.); a full spectroscopic characterization is given in the original paper.[2]

14. The thiobutyric acid was prepared according to A. Fredga and H. Bauer, *Arkiv. Kemi.*, **2**, 115 (1951), as follows: A rapid stream of hydrogen sulphide is passed, with vigorous stirring at −30°, through 200 ml. of anhydrous pyridine contained in a four-necked, round-bottomed flask equipped with a sealed mechanical stirrer, a gas-inlet tube, a pressure-equalizing dropping funnel, and a thermometer. Over approximately 1 hour, 50 g. of 1-butyryl chloride is added dropwise to this solution. Then approximately 400 ml. of aqueous 5N sulfuric acid is added slowly until the pH is ~5. The organic acid, which separates as a yellow oil, is taken up in ether and dried over anhydrous sodium sulfate. After removal of the ether on a rotary evaporator, the product is distilled through a 14-cm. Vigreux column under a nitrogen atmosphere to yield 23.1–32.1 g. (43–65%, yield not optimized by the submitters) of thiobutyric acid as a colorless liquid, b.p. 119–121°.

15. Triethylamine was purchased from Fluka AG or J. T. Baker Chemical Company.

16. A colorless precipitate of triethylamine hydrobromide is formed immediately. The temperature rises to about 35°.

17. 3-Bromo-2-butanone was purchased from Fisher Scientific Company. The submitters prepared it according to J. R. Catch,

D. F. Elliott, D. H. Hey, and E. R. H. Jones, *J. Chem. Soc. London*, 272, 278 (1948) and checked its purity (>95%) by gas chromatography (Note 2).

18. Silica gel (70–230 mesh ASTM) purchased from E. Merck & Company, Inc., Darmstadt, Germany was used in a 2.5-cm. diameter column.

19. Gas chromatographic analysis (Note 2) indicated <2% impurities. Infrared (neat) cm.$^{-1}$: 1720, 1695.

20. The submitters obtained a similar yield on ten times the scale.

21. The absence of water in the lithium bromide is of great importance. Traces of water lower the yield of product by 10–20%. LiBr·2 H$_2$O (purchased from E. Merck & Company, Inc., Darmstadt or City Chemical Corporation) was dissolved three times in anhydrous acetonitrile–benzene (1:1), and the solvents removed each time on a rotary evaporator. The lithium bromide was dried under high vacuum at 100° for 1 hour, ground to a fine powder with a mortar and pestle while still warm, and again dried at 100°, as above, for 3 hours.

22. Acetonitrile was purchased from Fluka AG or J. T. Baker Chemical Company and was distilled from potassium carbonate immediately prior to use.

23. The reaction is followed best by gas chromatographic analysis (Note 2). Traces of water seem to slow down the rate of the reaction.

24. By gas chromatographic analysis (Note 2) the product is >98% pure. In the literature,[2] a full spectroscopic characterization is given. Infrared (neat) cm.$^{-1}$: 1725, 1700, 1600, 1360.

3. Discussion

This procedure illustrates a broadly applicable method which is essentially that of Roth, Dubs, Götschi, and Eschenmoser,[2] for the synthesis of enolizable β-dicarbonyl compounds. Although there are various methods for the preparation of β-dicarbonyl systems,[3] the scheme of sulfide contraction widens the spectrum of available methods. The procedure can also be utilized in the synthesis of aza and diaza analogs of β-dicarbonyl systems. Eschenmoser[2] has utilized the method to produce vinylogous amides and amidines in connection with the total synthesis of corrins and vitamin B$_{12}$.[4]

S-Alkylation of a thiocarboxylic acid with an α-halogenated carbonyl compound gives a thiol ester in which the two carbons to be connected are linked *via* a sulfur bridge (see the scheme below). Enolization and formation of the episulfide creates the desired carbon-carbon bond. Removal of atomic sulfur by a thiophile, either a phosphine or a

phosphite, liberates the β-dicarbonyl compound. The addition of base is necessary in most cases; however, in the vinylogous amidine systems[4] electrophilic catalysis was employed. Normally a tertiary alkoxide is utilized to perform the contraction. The addition of anhydrous lithium bromide or lithium perchlorate allows the reaction to proceed with the use of a tertiary amine as the base. Presumably, the lithium salts complex with the carbonyl groups to enhance the enolization and/or contraction step.

This procedure also incorporates the use of bis(3-dimethylamino-propyl)phenylphosphine as a combined amine–phosphine reagent. The merits of using this basic phosphine as opposed to a tertiary amine and a phosphine lies in the ease of workup. Excess phosphine and phosphine sulfide can be removed by extraction with aqueous dilute acid.

Since the new carbon-carbon bond is formed intramolecularly in the sulfide extrusion method, its main potential is to be seen in cases where intermolecular condensations fail.[5,6]

1. SOCAR AG, Dübendorf, Switzerland.
2. M. Roth, P. Dubs, E. Götschi, and A. Eschenmoser, *Helv. Chim. Acta*, **54**, 710 (1971).
3. For a review see H. O. House, "Modern Synthetic Reactions," W. A. Benjamin, Menlo Park, California, 1972, pp. 734–785.
4. (a) Y. Yamada, D. Miljkovic, P. Wehrli, B. Golding, P. Loeliger, R. Keese, K. Müller, and A. Eschenmoser, *Angew. Chem.*, **81**, 301 (1969), *Angew. Chem., Int. Ed. Engl.*, **8**, 343 (1969); A. Eschenmoser, *Quart. Rev. Chem. Soc.*, **24**, 366 (1970); A. Eschenmoser, Special Lectures, 23rd International Congress of Pure and Applied Chemistry, Vol. II, Butterworth and Company, Ltd., London, 1971, pp. 69–106. (b) R. B. Woodward, *Pure Appl. Chem.*, **17**, 519 (1968); *Pure Appl. Chem.*, **25**, 283 (1971); *Pure Appl. Chem.*, **33**, 145 (1973).
5. I. Felner and K. Schenker, *Helv. Chim. Acta*, **53**, 754 (1970).
6. A. Gossauer and W. Hirsch, *Tetrahedron Lett.*, 1451 (1973).

IODOMETHANE

WARNING

Iodomethane in high concentrations for short periods or in low concentrations for long periods can cause serious toxic effects in the central nervous system. Accordingly the American Conference of Governmental Industrial Hygienists[1] has set 5 p.p.m. as the highest average concentration in air to which workers should be exposed for long periods. This low level cannot be detected by smell. Therefore the preparation and use of iodomethane should always be performed in a well-ventilated fume hood. Since liquid iodomethane can be absorbed through the skin, care should be taken not to expose hands or other skin to liquid iodomethane.

1. American Conference of Governmental Industrial Hygienists (ACGIH), "Documentation of Threshold Limit Values," 3rd ed., Cincinnati, Ohio, 1971, p. 166.

AUTHOR INDEX

This index comprises the names of contributors to Volume 55 only. For previous volumes see Collective Volumes 1 through 4 and Volume 54.

SUBJECT INDEX

This index comprises material from Volume 55 only; for subjects of previous volumes see Collective Volumes 1-5 and Volume 54.

The index consists of two parts. Part I contains entries referring to the conventional names of compounds as they appear in the preparations of this volume followed by the systematic names in brackets, if the nomenclature differs. The bracketed names conform to the systematic nomenclature adopted by the Chemical Abstracts Service starting with Chemical Abstracts Volume 76 (1972). Part II of the index contains entries of the subjects of this volume with compound names in a reversed order of Part I.

Entries in CAPITAL LETTERS are used for the titles of individual preparations. Entries in ordinary type letters refer to principal products and major by-products, special reagents or intermediates (which may or may not be isolated), compounds mentioned in the text, Notes or Discussions as having been prepared by the method given, and apparatus in detail or illustrated by a figure.

Part I

Acetic acid, butyryl-, ethyl ester [Hexanoic acid, 3-oxo-, ethyl ester], **55**, 73, 75

Acetic acid, chloro-, *tert*-butyl ester [Acetic acid, chloro-, 1,1-dimethylethyl ester], **55**, 94

Acetic acid, cyano-, ethyl ester, **55**, 58, 60

Acetic acid, 3,4-dimethoxyphenyl-, **55**, 45, 46

Acetic acid, nitro-, dipotassium salt, **55**, 77, 78

Acetic acid, nitro-, METHYL ESTER, **55**, 77, 78

Acetic acid, trifluoro-, **55**, 70

Acetone, phenyl- [2-Propanone, 1-phenyl-], **55**, 25

Acetonitrile, diphenyl-, **55**, 94, 102

Acetonitrile, diphenyl-2-(1-ethoxyvinyl)- [Acetonitrile, diphenyl-2-(1-ethoxyethenyl)-], **55**, 102

Acetonitrile, phenyl-, **55**, 91, 94

Acetophenone, 4-chloro- [Ethanone, 1-(4-chlorophenyl)-], **55**, 40

Acetophenone, 4-chloro-, oxime [Ethanone, 1-(4-chlorophenyl)-, oxime], **55**, 39, 40

Acetyl chloride, α-*tert*-butyl-α-cyano- [Butyryl chloride, 2-cyano-3,3-dimethyl-], **55**, 38

Acetylene, butylthio- [Ethyne, butylthio-], **55**, 102

Acetylene, ethoxy- [Ethyne, ethoxy-], **55**, 102

Acetylene, phenyl- [Ethyne, phenyl-], **55**, 102

Acetylene [Ethyne], **55**, 100

ALDEHYDES, acetylenic [Aldehydes, ethynic], **55**, 52

aromatic, aromatic hydrocarbons from, **55**, 7

Alicyclic compounds, **55**, 61

Alkenes, **55**, 3, 103

Alkenyl alcohols, *(E)*- disubstituted, **55**, 66

ALKYL HALIDES, alkenes from, **55**, 103

C-ALKYLATION, phase transfer catalysis

Part II

Received during the period July 1, 1974 - June 30, 1975
and
subsequently accepted for checking

In accordance with a policy adopted by the Board of Editors, beginning with Volume 50 and further modified in the present Volume as noted in the Editor's Preface, procedures received by the Secretary during the year and subsequently accepted for checking by Organic Syntheses, will be made available for purchase at the price of $2 per procedure, prepaid, upon request to the Secretary:

> Dr. Wayland E. Noland, Secretary
> Organic Syntheses
> Department of Chemistry
> University of Minnesota
> Minneapolis, MN 55455

Payment must accompany the order, and should be made payable to Organic Syntheses, Inc. (not to the Secretary). Purchase orders not accompanied by payment will not be accepted. Procedures may be ordered by number and/or title from the list which follows.

It should be emphasized that the procedures which are being made available are unedited and have been reproduced just as they are first received from the submitters. There is no assurance that the procedures listed here will ultimately check in the form available, and some of them may be rejected for publication in Organic Syntheses during or after the checking process. For this reason, Organic Syntheses can provide no assurance whatsoever that the procedures will work as described, and offers no comment as to what safety hazards may be involved. Consequently, more than usual caution should be employed in following the directions in the procedures.

Organic Syntheses welcomes, on a strictly voluntary basis, comments from persons who attempt to carry out the procedures. For this purpose, a Checker's Report form will be mailed out with each unchecked procedure ordered. Procedures which have been checked by or under the supervision of a member of the Board of Editors will continue to be published in the volumes of Organic Syntheses, as in the past. It is anticipated that many of the procedures in the list will be published (often in revised form) in Organic Syntheses in future volumes.

-Wayland E. Noland

4-*tert*-Butylcyclohexanone *via* A Polymeric Carbodi-
imide

N. M. Weinshenker, C. M. Shen, and J. Y. Wong, Dynapol, 1454 Page
Mill Road, Palo Alto, CA 94304

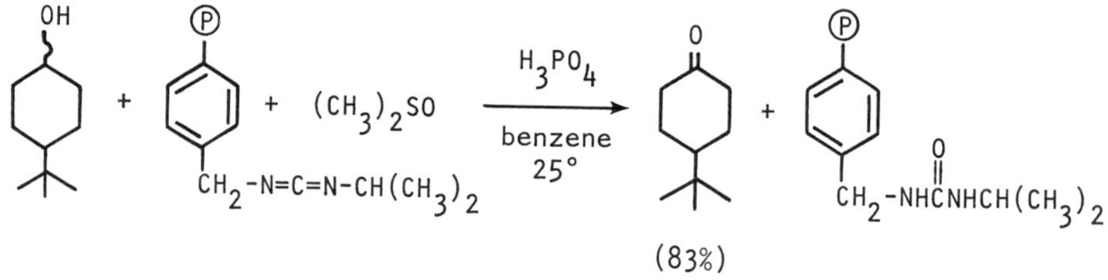

(83%)

1926   (Z)-3-Methyl-2-pentenoic Acid

J. F. Normant, G. Cahiez, C. Chuit, and J. Villieras, Laboratoire
de Chimie des Organoéléments, Université Paris VI, U.E.R. 56, Tour
44-45-4, Place Jussieu, 75005 Paris, France

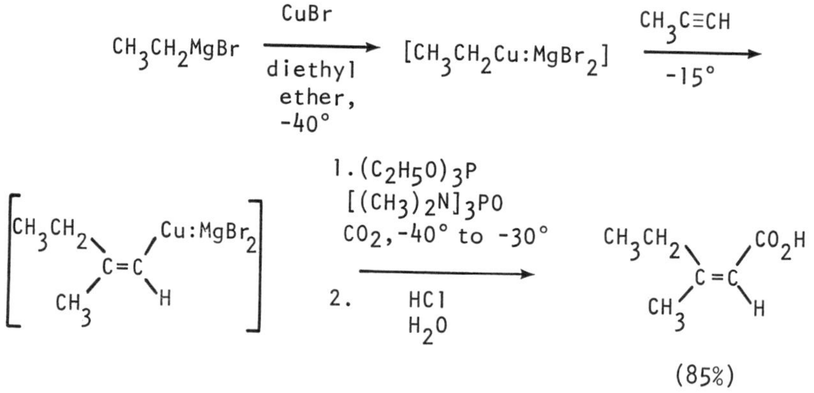

(85%)

1933   3-Phenyl-2-formyl-2H-azirine

A. Padwa, T. Blacklock, and A. Tremper, Department of Chemistry,

State University of New York at Buffalo, Buffalo, NY 14214

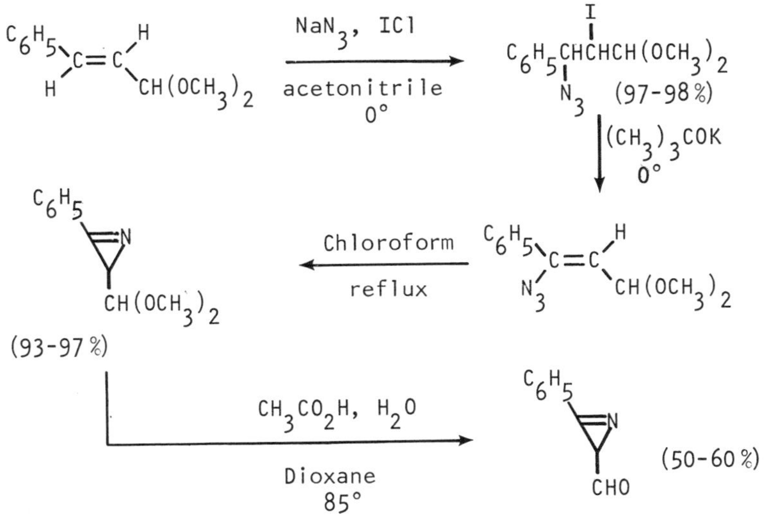

$$C_6H_5\underset{H}{\overset{H}{C}}=C\underset{CH(OCH_3)_2}{} \xrightarrow[\text{acetonitrile} \atop 0°]{\text{NaN}_3,\ \text{ICl}} C_6H_5\underset{N_3}{\overset{I}{CH}}CHCH(OCH_3)_2 \quad (97\text{-}98\%)$$

$$\downarrow \overset{(CH_3)_3COK}{0°}$$

$$\underset{\underset{(93\text{-}97\%)}{CH(OCH_3)_2}}{\overset{C_6H_5}{\triangle}N} \xleftarrow[\text{reflux}]{\text{Chloroform}} C_6H_5\overset{H}{\underset{N_3}{C}}=C\underset{CH(OCH_3)_2}{}$$

$$\downarrow \xrightarrow[\text{Dioxane} \atop 85°]{CH_3CO_2H,\ H_2O} \underset{CHO}{\overset{C_6H_5}{\triangle}N} \quad (50\text{-}60\%)$$

1934   18-Crown-6

G. W. Gokel, D. J. Cram, C. L. Liotta, H. P. Harris, and F. L. Cook,

Department of Chemistry, Pennsylvania State University, University

Park, PA 16802

$$H(OCH_2CH_2)_3OH\ +\ Cl(CH_2CH_2O)_2CH_2CH_2Cl \xrightarrow[\text{tetrahydro-} \atop \text{furan, reflux}]{KOH,\ H_2O}$$

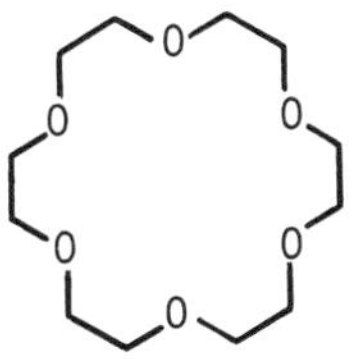

(40-50%)

Phenanthrene 9,10-Oxide

C. Cortez and R. G. Harvey, Ben May Laboratory, University of

Chicago, Chicago, IL 60637

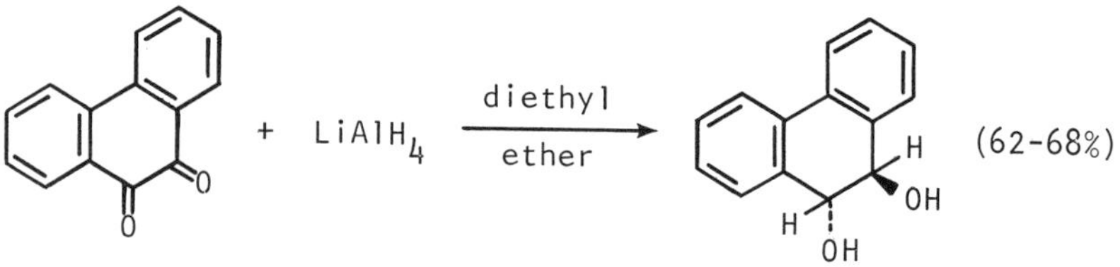

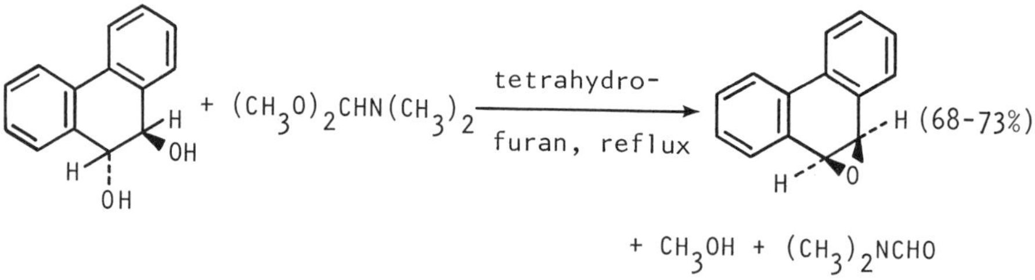

$+$ LiAlH$_4$ $\xrightarrow[\text{ether}]{\text{diethyl}}$ (62-68%)

$+$ (CH$_3$O)$_2$CHN(CH$_3$)$_2$ $\xrightarrow[\text{furan, reflux}]{\text{tetrahydro-}}$ (68-73%)

$+$ CH$_3$OH $+$ (CH$_3$)$_2$NCHO

1936   p-Toluenesulfonyldiazomethane

A. M. van Leusen and J. Strating, Department of Organic Chemistry,

University of Groningen, Zernikelaan, Groningen, The Netherlands

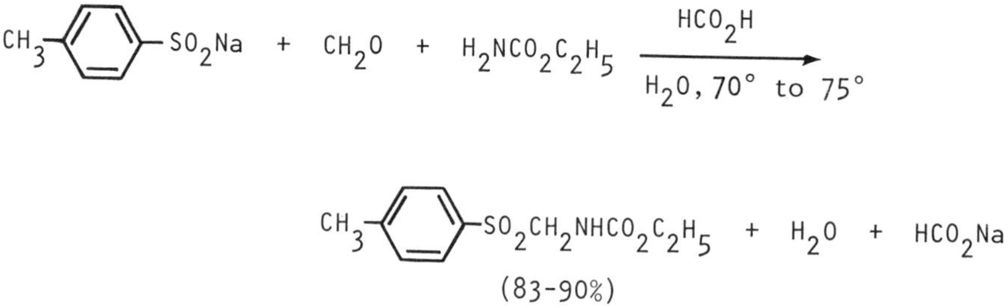

CH$_3$—⟨⟩—SO$_2$Na $+$ CH$_2$O $+$ H$_2$NCO$_2$C$_2$H$_5$ $\xrightarrow[\text{H}_2\text{O, 70° to 75°}]{\text{HCO}_2\text{H}}$

CH$_3$—⟨⟩—SO$_2$CH$_2$NHCO$_2$C$_2$H$_5$ $+$ H$_2$O $+$ HCO$_2$Na

(83-90%)

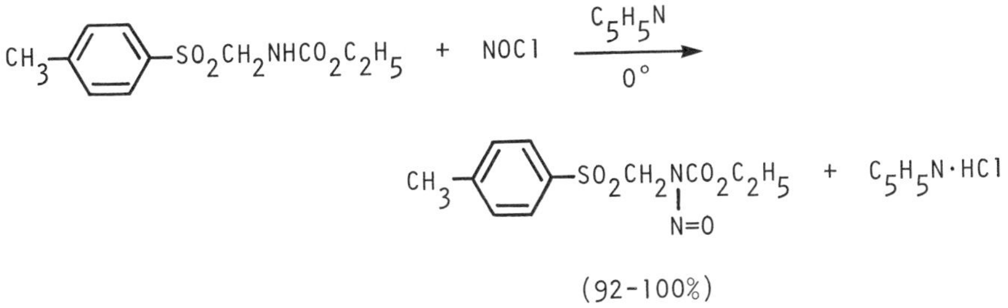

CH_3—⟨C6H4⟩—$SO_2CH_2NHCO_2C_2H_5$ + NOCl $\xrightarrow[0°]{C_5H_5N}$

CH_3—⟨C6H4⟩—$SO_2CH_2\underset{\underset{N=O}{|}}{N}CO_2C_2H_5$ + $C_5H_5N \cdot HCl$

(92-100%)

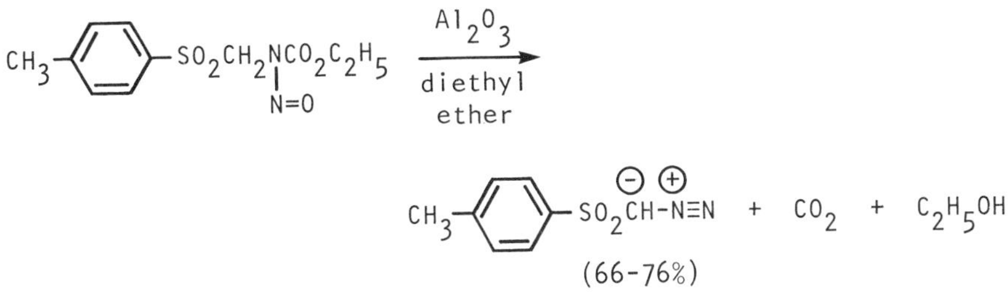

CH_3—⟨C6H4⟩—$SO_2CH_2\underset{\underset{N=O}{|}}{N}CO_2C_2H_5$ $\xrightarrow[\substack{diethyl \\ ether}]{Al_2O_3}$

CH_3—⟨C6H4⟩—$SO_2\overset{\ominus}{C}H-\overset{\oplus}{N}{\equiv}N$ + CO_2 + C_2H_5OH

(66-76%)

1937   Cyclopropenone

R. Breslow, J. Pecoraro, and T. Sugimoto, Department of Chemistry,
Columbia University, New York, NY   10027

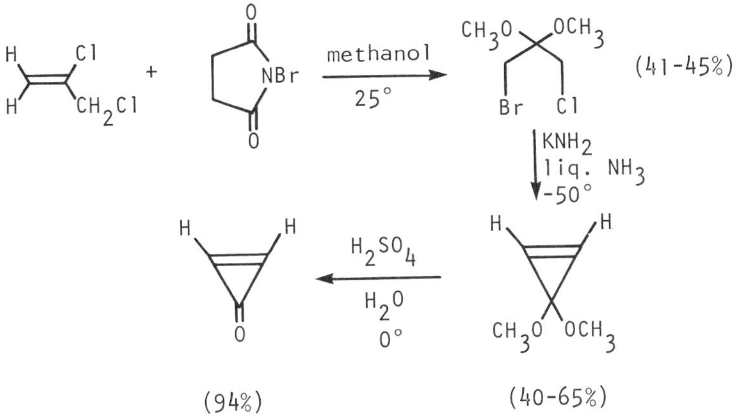

(41-45%)

KNH_2
liq. NH_3
-50°

H_2SO_4
H_2O
0°

(94%)   (40-65%)

1938   <u>3,3,6,6-Tetramethoxy-1,4-cyclohexadiene</u>

P. Margarétha and P. Tissot, Département de Chimie Organique,

Section de Chimie, Université de Genève, 30, quai Ernest-Ansermet,

1211 Genève 4, Switzerland

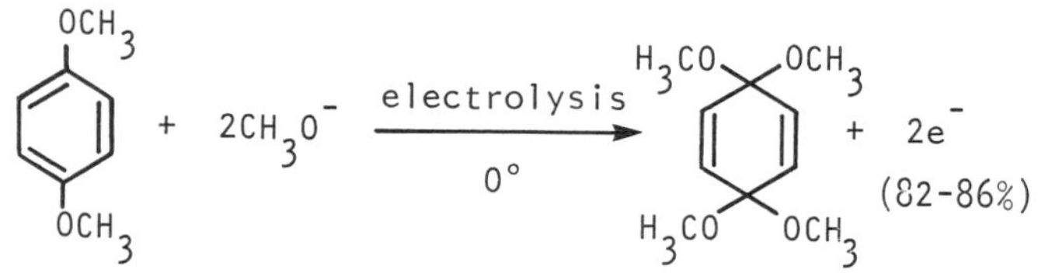

1939   <u>Oxidation with Bis(salicylidene)ethylenediiminocobalt</u>

             <u>(II) (Salcomine)</u>. Preparation and Use of Salcomine:

             <u>2,6-Diphenyl-p-benzoquinone</u>

C.R.H.I. de Jonge, H. J. Hageman, G. Hoentjen, and W. J. Mijs,

Akzo Research & Engineering Div., Velperweg 76, Postbus 60,

Arnhem, The Netherlands

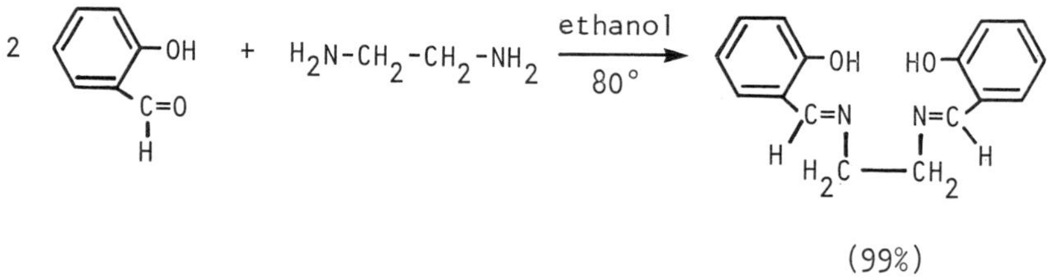

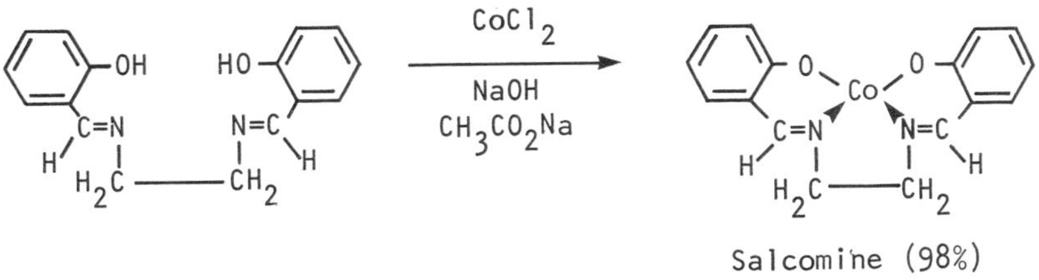

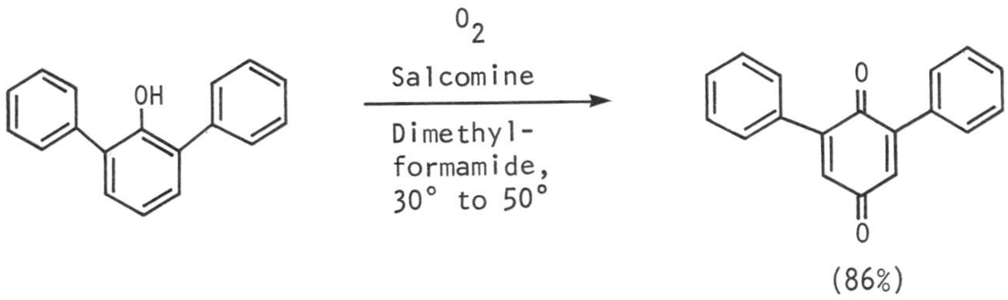

$$\xrightarrow[\substack{NaOH \\ CH_3CO_2Na}]{CoCl_2}$$

Salcomine (98%)

$$\xrightarrow[\substack{Dimethyl- \\ formamide, \\ 30° \text{ to } 50°}]{\substack{O_2 \\ Salcomine}}$$

(86%)

1940   Preparation of 1,2-Dimethylcyclobutenes by Reductive Ring Contraction of Sulfolanes: *cis*-7,8-Dimethyl-bicyclo[4.2.0]oct-7-ene

J. M. Photis and L. A. Paquette, Department of Chemistry, The Ohio State University, Columbus, OH 43210

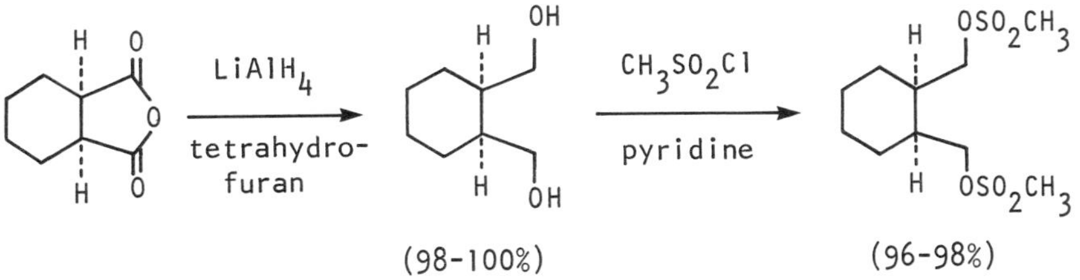

$$\xrightarrow[\substack{tetrahydro- \\ furan}]{LiAlH_4}$$

(98-100%)

$$\xrightarrow[pyridine]{CH_3SO_2Cl}$$

(96-98%)

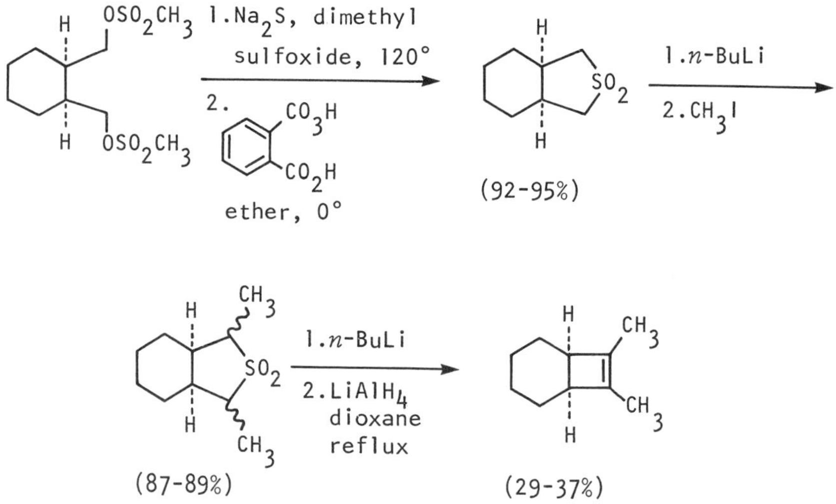

(92-95%)

(87-89%)   (29-37%)

1941   <u>p-Toluenesulfonylmethyl Isocyanide</u>

B. E. Hoogenboom, O. H. Oldenziel, and A. M. van Leusen, Department of Organic Chemistry, University of Groningen, Zernikelaan, Groningen, The Netherlands

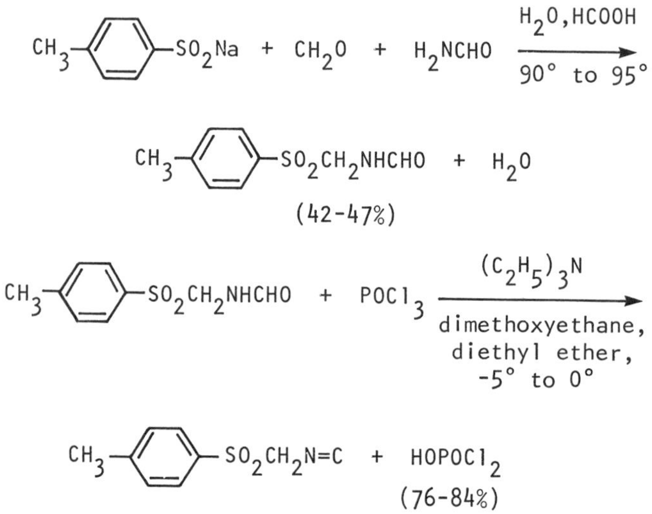

$$CH_3-\langle\rangle-SO_2Na \ + \ CH_2O \ + \ H_2NCHO \xrightarrow[90° \text{ to } 95°]{H_2O, HCOOH}$$

$$CH_3-\langle\rangle-SO_2CH_2NHCHO \ + \ H_2O$$

(42-47%)

$$CH_3-\langle\rangle-SO_2CH_2NHCHO \ + \ POCl_3 \xrightarrow[\substack{\text{dimethoxyethane,}\\ \text{diethyl ether,}\\ -5° \text{ to } 0°}]{(C_2H_5)_3N}$$

$$CH_3-\langle\rangle-SO_2CH_2N=C \ + \ HOPOCl_2$$

(76-84%)

1942   2-Adamantanecarbonitrile

O. H. Oldenziel, J. Wildeman, and A. M. van Leusen, Department of

Organic Chemistry, University of Groningen, Zernikelaan, Groningen,

The Netherlands

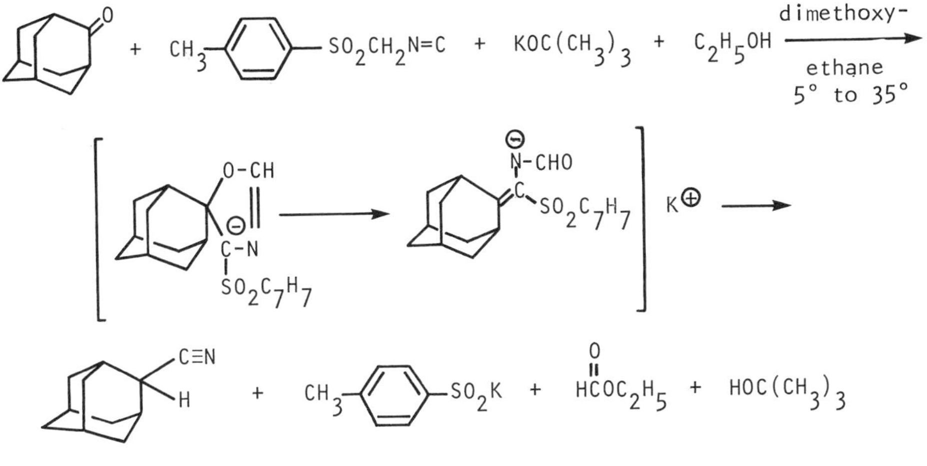

(84-90%)

1945   Diethylaminosulfur Trifluoride

W. J. Middleton and E. M. Bingham, Central Research and Development

Department, E. I. duPont deNemours and Company, Experimental

Station, Wilmington, DE 19898

$$(C_2H_5)_2N-Si(CH_3)_3 \; + \; SF_4 \longrightarrow (C_2H_5)_2NSF_3 \; + \; FSi(CH_3)_3$$

(80-90%)

1946   α-Fluoro-*p*-nitrotoluene

W. J. Middleton and E. M. Bingham, Central Research and Development
Department, E. I. duPont deNemours and Company, Experimental
Station, Wilmington, DE 19898

$$O_2N-\!\!\!\bigcirc\!\!\!-CH_2OH \;+\; (C_2H_5)_2NSF_3 \xrightarrow[10° \text{ to } 25°]{}$$

$$O_2N-\!\!\!\bigcirc\!\!\!-CH_2F \;+\; (C_2H_5)_2NSOF \;+\; HF$$

(65-70%)

1947   5,6-Dihydro-2-pyrone and 2-Pyrone

M. Nakagawa, J. Saegusa, M. Tonozuka, M. Obi, M. Kiuchi, T. Hino,
and Y. Ban, Faculty of Pharmaceutical Sciences, Hokkaido
University, Sapporo, 060 Japan

(Z),(Z)-2,6,10-Trimethyldodeca-2,6,10-triene

E. J. Corey, Department of Chemistry, Harvard University,

Cambridge, MA 02138, and K. Achiwa, Faculty of Pharamaceutical

Sciences, University of Tokyo, Hongo, Tokyo, Japan

$$C_5H_5N \xrightarrow[\substack{\text{dichloro-}\\\text{methane, } 0°}]{ClSO_3H} C_5H_5N \cdot SO_3$$

$$(78\%)$$

OH + $C_5H_5N \cdot SO_3$ $\xrightarrow[\substack{\text{furan, } 0°}]{\text{tetrahydra-}}$

$OSO_3^- C_5H_5\overset{+}{N}H$ $\xrightarrow[\substack{\text{tetrahydro-}\\\text{furan}\\0° \text{ to } 25°}]{LiAlH_4}$

$$(80\%)$$

Preparation of γ-Ketoesters from Aldehydes *via*
      Diethyl Acylsuccinates

P. A. Wehrli and V. Chu, Hoffmann-La Roche Inc., Chemical Research
Department, Nutley, NJ   07110

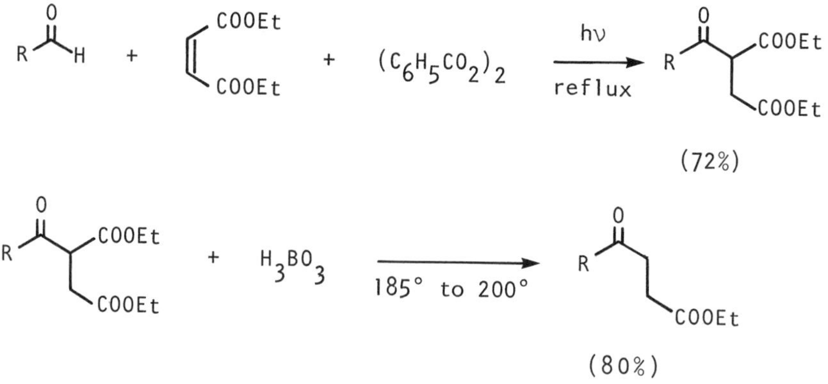

(72%)

(80%)

2-Methylcyclopentane-1,3-dione

U. Hengartner and V. Chu, Hoffmann-La Roche Inc., Chemical Research
Department, Nutley, NJ   07110

(71%)

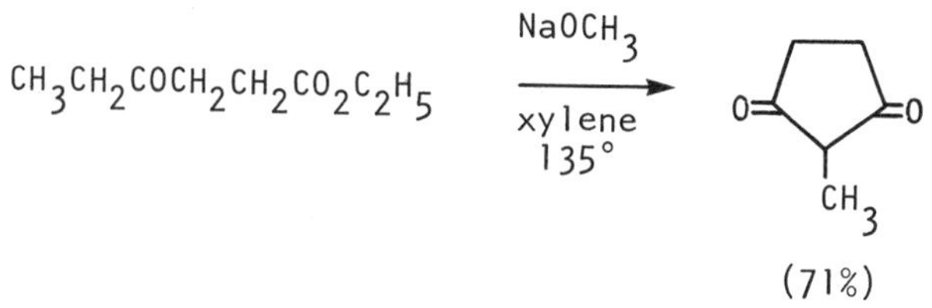

1955   Conversion of Nitro to Carbonyl by Ozonolysis of

Nitronates: 2,5-Heptanedione

J. McMurry and J. Melton, Thimann Laboratories, University of

California - Santa Cruz, CA 95064

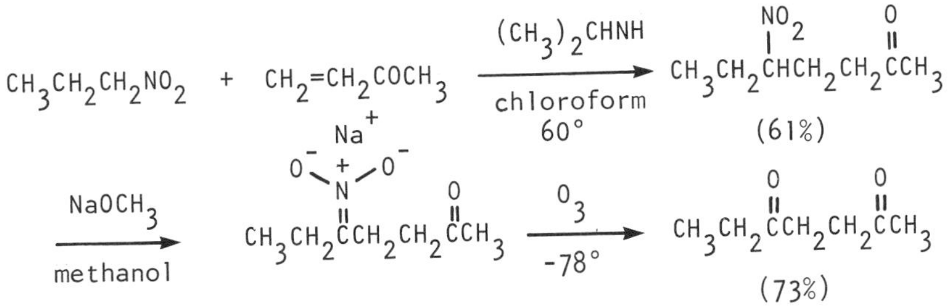

$$CH_3CH_2CH_2NO_2 \; + \; CH_2=CH_2COCH_3 \; \xrightarrow[\substack{\text{chloroform} \\ 60°}]{(CH_3)_2CHNH} \; CH_3CH_2\overset{\overset{\displaystyle NO_2}{|}}{C}HCH_2CH_2\overset{\overset{\displaystyle O}{\|}}{C}CH_3$$

(61%)

$$\xrightarrow[\text{methanol}]{NaOCH_3} \; CH_3CH_2\overset{\overset{\displaystyle O^-\overset{+}{\underset{\|}{N}}O^-}{}}{C}CH_2CH_2\overset{\overset{\displaystyle O}{\|}}{C}CH_3 \; \xrightarrow[-78°]{O_3} \; CH_3CH_2\overset{\overset{\displaystyle O}{\|}}{C}CH_2CH_2\overset{\overset{\displaystyle O}{\|}}{C}CH_3$$

(73%)

1956   Oxidation of Alcohol by DMS-NCS-TEA

E. J. Corey, Department of Chemistry, Harvard University, Cambridge,

MA 02138, and C. U. Kim and P. F. Misco, Antibiotic Chemical

Research, Bristol Laboratories, P. O. Box 657, Syracuse, NY 13201

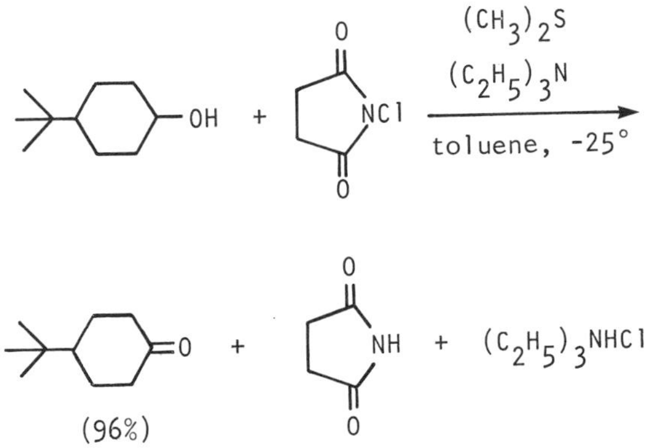

Phosphine-Nickel Complex-catalyzed Cross-coupling of Grignard Reagents with Aryl and Alkenyl Halides. o-Di-n-Butylbenzene

M. Kumada, K. Tamao, and K. Sumitani, Department of Synthetic Chemistry, Faculty of Engineering, Kyoto University, Yoshida, Kyoto 606, Japan

$$CH_3(CH_2)_3Br + Mg \xrightarrow[\text{ether}]{\text{diethyl}} CH_3(CH_2)_3MgBr$$

$$+ \ 2CH_3(CH_2)_3MgBr \xrightarrow[\substack{\text{diethyl ether,}\\ 0°, \text{ then reflux}}]{\sim 10^{-3} \text{ eq. } [Ni(dppp)Cl_2]}$$

(79-83%) $+ \ 2MgBrCl$

$$dppp = (C_6H_5)_2P(CH_2)_3P(C_6H_5)_2$$

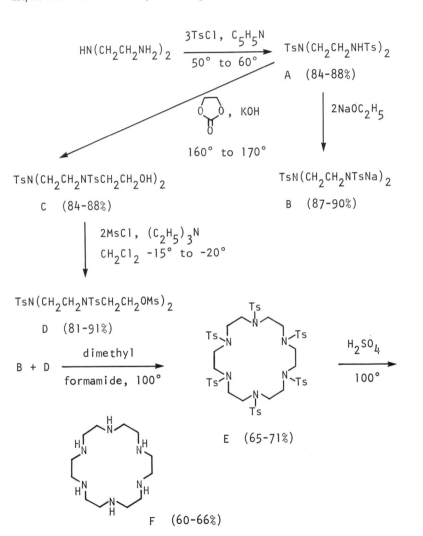

1958   1,4,7,10,13,16-Hexaazacyclooctadecane

T. J. Atkins, J. E. Richman, and W. F. Oettle, Central Research
and Development Department, E. I. duPont deNemours and Company,
Experimental Station, Wilmington, DE   19898

$$HN(CH_2CH_2NH_2)_2 \xrightarrow[50° \text{ to } 60°]{3TsCl, \ C_5H_5N} TsN(CH_2CH_2NHTs)_2$$

A   (84-88%)

(with diagonal arrow) $\xrightarrow{\overset{O}{\underset{O \quad O}{\bigcirc}}, \ KOH}$ 160° to 170°

$\downarrow 2NaOC_2H_5$

$TsN(CH_2CH_2NTsCH_2CH_2OH)_2$

C   (84-88%)

$TsN(CH_2CH_2NTsNa)_2$

B   (87-90%)

$\downarrow$ 2MsCl, $(C_2H_5)_3N$
CH_2Cl_2 -15° to -20°

$TsN(CH_2CH_2NTsCH_2CH_2OMs)_2$

D   (81-91%)

$B + D \xrightarrow[\text{formamide, } 100°]{\text{dimethyl}}$ E $\xrightarrow[100°]{H_2SO_4}$ F

E   (65-71%)

F   (60-66%)

Methyl 5-Ethoxycarbonyl-6-methyl-2(1*H*)-pyridon-4-

          ethanoate

T. A. Bryson and T. M. Dolak, Department of Chemistry, University

of South Carolina, Columbia, SC 29208

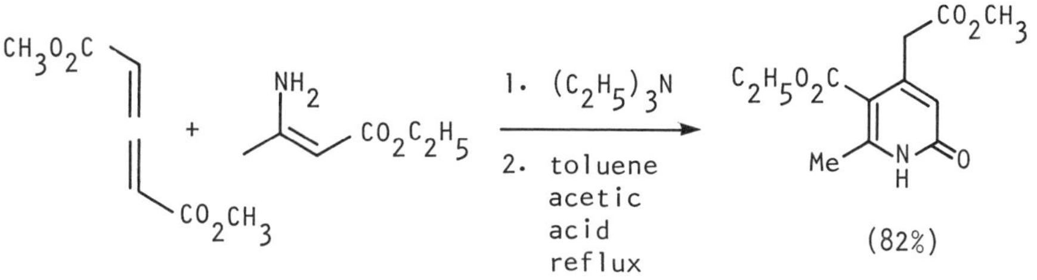

(82%)

Cyclopentenones from α,α'-Dibromoketones and En-

          amines: 2,5-Dimethyl-3-phenyl-2-cyclopenten-1-one

R. Noyori, K. Yokoyama, and Y. Hayakawa, Department of Chemistry,

Faculty of Science, Nagoya University, Chikusa, Nagoya 464, Japan

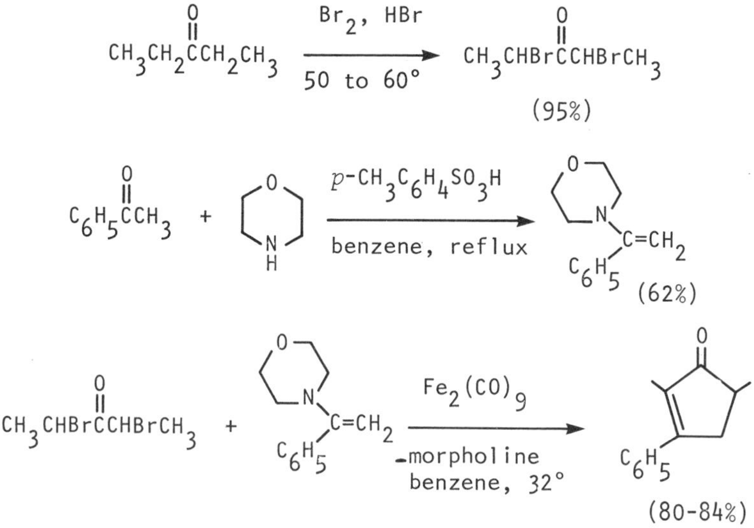

(95%)

(62%)

(80-84%)

<u>Allylically Transposed Amines from Allylic Alcohols:</u>

<u>3-Amino-3,7-dimethyl-1,6-octadiene (Linalylamine)</u>

L. A. Clizbe and L. E. Overman, Department of Chemistry, University

of California - Irvine, CA 92664

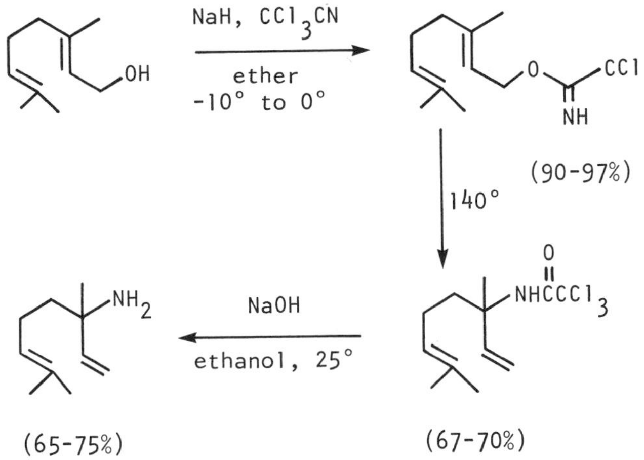

(90-97%)

140°

NaOH

ethanol, 25°

(65-75%)   (67-70%)

<u>2-(1-Acetyl-2-oxopropyl)benzoic Acid</u>

A. Bruggink, A. McKillop, and S. J. Ray, School of Chemical Sciences,

University of East Anglia, University Plain, Norwich, England, U.K.

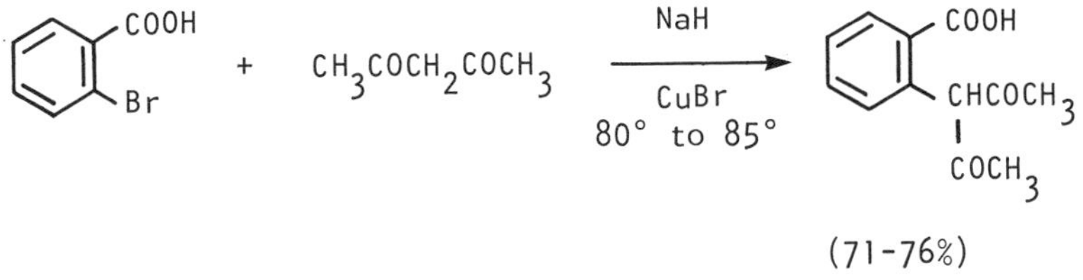

(71-76%)

1967   A Convenient Cyclization of ω-Bromocarboxylic Acids

to Macrolides: 11-Undecanolide

C. Galli and L. Mandolini, Centro di Studio sui Meccanismi di

Reazione del Consiglio Nazionale delle Ricerche c/o Instituto

Chimico dell'Università di Roma, 00185   Rome, Italy

$$Br(CH_2)_{10}CO_2H \xrightarrow[\substack{\text{dimethyl sulfoxide} \\ 100°}]{K_2CO_3} (CH_2)_{10}\overset{\overset{O}{\|}}{C}=O$$

$$(78-80\%)$$

1969   A New General Synthesis of 3-Alkylalk-1-ynes:

3-Ethylhex-1-yne

A. J. Quillinan and F. Scheinmann, Department of Chemistry and

Applied Chemistry, University of Salford, Salford M5 4WT,

England, U.K.

$$CH_3(CH_2)_2CH_2C\equiv CH \xrightarrow[\substack{\text{pentane} \\ -20° \text{ to } 10°}]{n\text{-BuLi}} [CH_3(CH_2)_2\overset{\overset{Li}{|}}{C}H-C\equiv CLi]$$

$$\xrightarrow[\substack{2. \text{ aqueous HCl} \\ 0° \text{ to } 5°}]{1. \ CH_3CH_2Br} CH_3(CH_2)_2\overset{\overset{CH_3CH_2}{|}}{C}H-C\equiv CH$$

$$(85\%)$$

WITHDRAWN
UNIV OF MOUNT UNION LIBRARY